Cutting a cake, dividing up the property in an estate, determining the borders in an international dispute – such problems of fair division are ubiquitous. *Fair division* treats all these problems and many more through a rigorous analysis of a variety of procedures for allocating goods (or "bads" like chores), or deciding who wins on what issues, when there are disputes. Starting with an analysis of the well-known cake-cutting procedure, "I cut, you choose," the authors show how it has been adapted in a number of fields and then analyze fair-division procedures applicable to situations in which there are more than two parties, or there is more than one good to be divided. In particular, they focus on procedures which provide "envy-free" allocations, in which everybody thinks he or she has received the largest portion and hence does not envy anybody else. They also discuss the fairness of different auction and election procedures.

Fair division

Fair division

From cake-cutting to dispute resolution

Steven J. Brams
Department of Politics
New York University
New York, NY 10003

and

Alan D. Taylor
Department of Mathematics
Union College
Schenectady, NY 12308

Published by the Press Syndicate of the University of Cambridge
The Pitt Building, Trumpington Street, Cambridge CB2 1RP
40 West 20th Street, New York, NY 10011–4211, USA
10 Stamford Road, Oakleigh, Melbourne 3166, Australia

First published 1996
Reprinted 1996

Printed in Great Britain at Athenæum Press Ltd, Gateshead, Tyne & Wear

A catalogue record for this book is available from the British Library

Library of Congress cataloguing in publication data

Brams, Steven J.
Fair division: from cake-cutting to dispute resolution /
Steven J. Brams and Alan D. Taylor.
p. cm.
Includes bibliographical references.
ISBN 0 521 55390 3 (hbk.) – ISBN 0 521 55644 9 (pbk.)
1. Conflict management. 2. Negotiation. 3. Fairness.
4. Game theory. I. Taylor, Alan D., 1947– . II. Title.
HM136.B7317 1996
303.6'9–dc20 95–16488 CIP

ISBN 0 521 55390 3 hardback
ISBN 0 521 55644 9 paperback

WD

"The distribution of the world's goods owes little to virtue . . . There must be better games. If I were to select a research problem without regard to scientific feasibility, it would be that of finding out how to persuade human beings to design and play games that all can win."

Herbert A. Simon (Simon, 1991, pp. 365–66)

"It seems that a lion, a fox, and an ass participated in a joint hunt. On request, the ass divides the kill into three equal shares and invites the others to choose. Enraged, the lion eats the ass, then asks the fox to make the division. The fox piles all the kill into one great heap except for one tiny morsel. Delighted at this division, the lion asks, 'Who has taught you, my very excellent fellow, the art of division?' to which the fox replies, 'I learnt it from the ass, by witnessing his fate.'"

One of Aesop's fables, as given in Lowry (1987, p. 130)

"Free, open-minded, and absolutely impartial adjustment of all colonial claims."

Point 5 of Woodrow Wilson's Fourteen Points, 1918

"Darling, it is appalling, those three ignorant and irresponsible men [Woodrow Wilson, Lloyd George, and Georges Clemenceau] cutting Asia Minor to bits as if they were dividing a cake."

Letter of Harold Nicolson to his wife, Vita Sackville-West, May 4, 1919 (Nicolson, 1992, p. 83; excerpted in Moynihan, 1993, p. 102)

"Gimme the Plaza, the jet and the $150 million, too."

Headline, *New York Post*, February 13, 1990, reporting Ivana Trump's divorce settlement demands of husband Donald

Contents

Figures

Tables

Acknowledgments

We gratefully acknowledge the valuable contributions of Julius B. Barbanel, Fred Galvin, D. Marc Kilgour, and William S. Zwicker to our thinking on different parts of this book. Specific observations and comments by A. M. Fink, David Gale, Sergiu Hart, Theodore P. Hill, William F. Lucas, Walter R. Stromquist, William A. Webb, Stephen J. Willson, and Douglas R. Woodall also proved helpful, as did a careful reading of an earlier version of the book manuscript by Julie C. Brams, Danny Kleinman, Hervé Moulin, Peter Landweber, and anonymous referees. In addition, we have much benefited from conversations and correspondence with the following people: Ethan Akin, Robert T. Clemen, John H. Conway, Terry Daniel, Morton D. Davis, Alfred de Grazia, Karl Dunz, Shimon Even, Peter C. Fishburn, Martin Gardner, Richard K. Guy, Randall G. Holcombe, Jerzy Legut, Joseph Malkevitch, Stephen B. Maurer, Samuel Merrill, III, Dominic Olivastro, Barry O'Neill, Philip Reynolds, J. H. Riejnierse, Steven M. Slutsky, Jordan Howard Sobel, Rafael Tenorio, William Thomson, Hal R. Varian, Markku Verkama, Charles Wilson, H. Peyton Young, and Daozhi Zeng. We also thank anonymous referees for many helpful suggestions, Jennifer A. Bell and Katri Saari for valuable research assistance, and John Haslam and Anne Rix of Cambridge University Press for, respectively, their strong support of the project and excellent editorial assistance.

Steven J. Brams is pleased to acknowledge the assistance of the C. V. Starr Center for Applied Economics at New York University, and Alan D. Taylor the assistance of the National Science Foundation under grant DMS-9101830.

Introduction

The problem of fair division is as old as the hills, but our approach to this problem is new. It involves

- setting forth explicit *criteria*, or *properties*, that characterize different notions of fairness;
- providing step-by-step *procedures*, or *algorithms*, for obtaining a fair division of goods or, alternatively, preferred positions on a set of issues in negotiations; and
- illustrating these algorithms with *applications* to real-life situations.

Our three-pronged focus on properties, algorithms, and applications pulls together work scattered in different disciplines. Thus, philosophers have devoted major attention to explicating the meaning of fairness, and related concepts like justice and equity, and spelling out some of their implications. Theoretical economists have postulated abstract properties that any fair-division scheme should satisfy, but their demonstration of the existence of such a scheme usually says little about how, *constructively* (i.e., by the use of an algorithm or step-by-step procedure), to produce a fair division. Mathematicians have proved nonconstructive existence results as well, but they have been more interested than economists in also developing algorithms that actually produce a fair division.[1]

Political scientists, sociologists, and applied economists have taken a more empirical tack, seeking to determine conditions under which fairness, or departures from it, occur in the world and what consequences they have for people and institutions. Psychologists have especially paid heed to how *perceptions* of fairness impinge on people's attitudes and affect their behavior.

Our approach to fair division is distinctive not only in combining properties, algorithms, and applications but also in elevating the property of "envy-freeness," and procedures that generate envy-free allocations, to a central place

[1] In 1946, mathematician Hermann Weyl, at a conference at Princeton University, put it this way: "When you can settle a question by explicit construction, be not satisfied with purely existential arguments" (quoted in Maurer, 1985, p. 2).

in the study of fair division. Roughly speaking, an *envy-free division* is one in which every person thinks he or she received the largest or most valuable portion of something – based on his or her own valuation – and hence does not envy anyone else. Although the concept of envy-freeness has been used in the mathematics literature on fair division for almost forty years (Gamow and Stern, 1958) and in the economics literature for almost thirty years (Foley, 1967), only recently have several algorithms been developed that guarantee envy-freeness in a wide variety of situations.

This guarantee has a specific meaning in game theory – namely, that if you adhere to the algorithm, the strategies that the other players select cannot prevent you from obtaining a portion that you think is the largest or most valuable. But while each player may obtain, say, a favorite piece of a cake, not all our algorithms ensure that the players could not have done better with an entirely different division of the same cake.[2]

The mathematics literature has particularly emphasized the problem of cake division, which is to say that most of the results are measure-theoretic or combinatorial in nature. Generally speaking, this literature assumes that the value of one's allocation is simply the sum of the values of the individual pieces one receives. Moving beyond the metaphor of cake, we shall explore applications of our algorithms to the allocation of less divisible goods in such areas as divorce settlement, estate division, and treaty negotiation, where again values are considered additive. We also extend our analysis to the study of institutions not conventionally associated with fair division, such as auctions and elections.

In the case of auctions, we propose that they be carried out in two stages, which ameliorates certain informational problems that plague not only one-stage auctions – inducing the so-called winner's curse – but also a classic bargaining problem, "divide-the-dollar." In the case of elections, we analyze four extant systems of proportional representation and then propose a new system that allows for the election of candidates representing two different characteristics of the electorate.

All the results in this book are built on rigorous mathematical foundations,

[2] That is, envy-free divisions can fail to be Pareto-optimal or Pareto-efficient – for simplicity, we shall use only the term *efficient* – because everybody may be able to obtain a bigger piece with another division. But if the price you pay for obtaining a bigger piece in an efficient division is to envy somebody who you think got still more than you did, then it may not be worthwhile. In fact, as we will show later, there are contexts in which envy-freeness and efficiency are incompatible, so you may have to make a choice between these two properties in a fair-division procedure. But not always: a two-person point-allocation scheme for divisible goods that we analyze, called "Adjusted Winner," is envy-free, efficient, and *equitable* – that is, each player "wins" by the same amount over 50 percent . However, this procedure is vulnerable, in theory, to *strategic manipulation* – caused by the false (or dishonest) point assignments of players – but in practice it is probably quite robust.

but we have tried to keep the presentation sufficiently informal, especially through the use of examples, so that the reader who wishes to eschew technical details will feel no lack of continuity. In particular, nontechnical readers – including social scientists, business people, lawyers, and others interested in issues of fairness, justice, and equity – should be able to read most of this book, especially the beginning, without difficulty. If they do get bogged down, we suggest that they skip over the more arcane material and focus instead on the concrete cases we use to illustrate the theory and procedures.

Fortunately, many of the fair-division schemes we discuss, including some discovered only in the last few years, have very intuitive interpretations. Because key features of some of these procedures can be described visually – for example, by a "moving knife" – we make liberal use of diagrams to elucidate their analysis.

On the other hand, the reader wishing to see the arguments which underlie the various claims we make will find virtually all of them here; only the more technical arguments in four of the eleven chapters (chapters 4, 7, 8, and 9) are put in appendices. At no point in the text do we make use of any symbolism or formal mathematical analysis beyond what is found in high school algebra.

To be sure, some of the more subtle procedures we discuss are founded on rather sophisticated logical arguments and, consequently, are not transparent. Although these arguments may require considerable thought and reflection to digest, we believe that the potential payoff from understanding them is substantial.

A thorough understanding of the reasoning underlying a procedure, for example, may suggest emendations in it that solve a significant new problem.[3] For example, the dual problem of dividing a "bad," such as household chores, so that each party thinks it gets the fewest, has recently been attacked. Its solution is applicable to such real-world allocation problems as burden-sharing in alliances, which states presumably want to minimize.

We have three aims in this book:

(1) to present the latest constructive results on fair-division procedures that have, in our view, rejuvenated theoretical study in the field;
(2) to illustrate the application of these procedures to important practical problems; and
(3) to trace out some of the history of fair division, including who discovered what.

We begin with procedures that guarantee *proportionality*, or shares of at least $1/n$ to each of the n players. (We assume in the beginning only two players; throughout the book we usually proceed from two to three or four to

[3] As Arrow (1992, p. 50) put it, "research should have not only results but pointers toward the incomplete."

the general case of *n* players). Several algorithms exist for this purpose, which we shall describe and link to real-life cases, but they are not envy-free: obtaining at least $1/n$ does not imply that a player will not envy someone else if that other player's share is perceived to be larger.

We next describe newer theoretical schemes, several found only in the 1990s, that guarantee envy-freeness. Some involve cake-cutting algorithms, which are most applicable to the division of divisible goods or issues but can, on occasion, be adapted to allow for indivisibilities. Others involve point-allocation or bidding schemes, which are more applicable to the division of indivisible goods or issues.

Because the latter schemes tend to be more practical and may guarantee efficiency and equitability as well (see note 2), we stress their application to a broad range of bargaining and fair-division problems. They and several of the other proportional and envy-free procedures also allow for *entitlements*, whereby some players have greater *a priori* claims than others that one may want reflected in the fair division. We will show how less entitled players can still be envy-free in a natural extension of the concept of envy-freeness.

A *leitmotif* of this book is the search for procedures that quench the flames of envy, which Webster defines as a "painful or resentful awareness of an advantage enjoyed by another joined with a desire to possess the same advantage."[4] Eliminating or at least minimizing envy has not, in our view, been sufficiently emphasized in attempts to solve a host of distributional problems involving both tangible goods (e.g., property in divorce settlements) and less tangible issues (e.g., sovereignty in international disputes), though a recent survey claims envy-freeness has become "absolutely central . . . in the literature on normative economics" and finds "a new surge of interest" in the subject (Arnsperger, 1994, p. 155).[5]

But efficiency and equity are also important desiderata. Because one generally cannot get everything – even theoretically – in a constructive fair-division scheme, tradeoffs are inevitable.

Nevertheless, new and promising results in the field, with clear practical ramifications, make us optimistic. In particular, we believe our adjusted-winner procedure, which we apply to the Panama Canal treaty negotiations in the 1970s, divorce settlements, and estate division, is sufficiently simple and practical that it deserves to be tried out in real-life situations. Similarly, the new auction and election procedures that we analyze help those who most value

[4] Social, political, psychological, and philosophical perspectives on envy can be found in, respectively, Schoeck (1969), de la Mora (1987), Salovey (1991), and Ben Ze'ev (1992).

[5] To be sure, a person may do better cultivating envy, as author Mark Twain understood when he had Tom Sawyer turn the tables on the boys who laughed at him for whitewashing a fence: when they, out of envy, became persuaded that the job was fun and wanted to take it over, Tom had the last laugh.

items, or who are most in need of representation, gain what they desire. They should, we believe, be considered serious candidates for adoption in place of certain procedures now in use.

The search for better procedures, which make the achievement of fairness not just an outcome but a process as well (Holcombe, 1983), will go on. What cannot wait is applying our knowledge, primitive as it is, to problems of fair division that cry out for better and more durable solutions in realistic settings.

1 Proportionality for $n = 2$

1.1 Biblical origins

In the Hebrew Bible, the issue of fairness is raised in some of the best-known narratives. Cain's raging jealousy and eventual murder of Abel is provoked by what he considered unfair treatment by God, who "paid heed" to Abel's offering but ignored his (Gen. 4:4).[1] Jacob, after doing seven years of service in return for Laban's beautiful daughter, Rachel, was told that his sacrifice was not sufficient and that he must instead marry Laban's older and plainer daughter, Leah, unless he did seven more years' service, which he regarded as not only breaking a contract but also flagrantly unfair. Fairness triumphed, however, when King Solomon proposed to divide a baby, claimed by two mothers, in two. When the true mother protested and offered the baby to the other mother (whose baby had died), the truth about the baby's maternity became apparent, and "all Israel . . . stood in awe of the king; for they saw that he possessed divine wisdom to execute justice" (1 Kings 3:28).

Solomon's proposed solution is the first explicit mention of fair division that we know of in recorded history. But it is, of course, no solution at all: Solomon had no intention of dividing the baby in two. Instead, his purpose was to set up a game between the two women, described in Brams (1980, 1990b), that would distinguish the mother from the impostor.

In pursuit of the truth, Solomon had foreseen the women's preferences:

- the mother's top priority would be saving her baby, even at the cost of losing him to the impostor;
- the impostor would disdain saving the baby and instead seek to curry Solomon's favor by not protesting his solution.

The game that Solomon set up between the women enabled him to interpret the strategies they chose, after announcing his own, as evidence of who was telling the truth and who was lying. In effect, he designed the rules of the game such that its play would distinguish truthfulness from mendacity.

[1] All Bible translations are from *The Torah: The Five Books of Moses* (1967) and *The Prophets* (1978).

The game "worked" – but only because the women did not discern Solomon's purpose. If the impostor, in particular, had been more perspicacious, then she, like the mother, would have protested, leaving Solomon in a quandary about who really was the mother.[2]

Solomon is venerated for his exemplary judgment, but there is no question that his setup involved deception. He should perhaps be extolled as much for his cunning in deceiving the impostor as for his probity in finding a just solution.

We do not applaud this kind of cunning when it is used by people, without the judicious temperament of a Solomon, to extract information for untoward purposes. Indeed, unscrupulous individuals may succeed not only in exploiting other persons through ingenious subterfuges, like Solomon's, but also in subverting agreements and undermining institutions.

Solomon's solution, which seems dazzling because it worked so well, is the exception in the Hebrew Bible. More common in both the Bible and the Talmud are discussions of how to divide up land or an estate, for which there are typically not brilliant or even compelling answers but different points of view.[3]

These usually reflect different notions of fairness, rooted in conflicting interests or principles. Our purpose is to find solutions that reconcile these interests or principles, insofar as possible, in a way that the participants themselves consider satisfactory. Furthermore, we insist that the participants be able to implement their own solution and not have to rely on an outside party, as is the case of arbitration.[4]

Not only do the solutions of the fair-division procedures not depend on the use of outside parties, but they also guarantee a certain "minimal" outcome to each player who uses the strategy, regardless of what the other players do. As we shall show, however, sometimes this minimal outcome can be improved upon through strategic maneuvers that, like Solomon's, capitalize on information that might be privileged, or at least not common knowledge.

Such strategizing, because it inevitably requires departing from the script of

[2] For formal treatments of King Solomon's dilemma, see Glazer and Ma (1989), Yang (1992), and Osborne and Rubinstein (1994, pp. 186–7, 190–2).

[3] Several classic solutions are reviewed in Young (1994, chapter 4).

[4] For an analysis of alternative arbitration schemes and their game-theoretic foundations, see Brams (1990, chapter 3) and Brams, Kilgour, and Merrill (1991) and references cited therein. An *arbitrator*, by definition, makes an arbitrary – though presumably fair – decision that the disputants must accept. Fair-division procedures, by contrast, leave to the disputants themselves what procedure, if any, they will adopt. (But once a procedure has been adopted, we assume, the decision reached under it is binding in the same manner that an arbitrator's choice is binding.) Although we rule out arbitrators in fair division, we commend the use of *mediators*, who may well serve as clarifiers and facilitators in disputes but who cannot, like arbitrators, dictate the settlement. Thus, for example, mediators may help the disputants define what the issues are, including what needs to be divided, but not decide what the division will be.

a fair-division procedure, is risky: it forces a player to give up the security of a guaranteed minimal outcome. Solomon's deception is no exception. If he had failed to outwit the impostor with his announcement, and she had protested too, Solomon might well have had to carry out his edict to divide the baby, which would have pleased nobody except, perhaps, the impostor.

Surely there were less radical solutions that Solomon could have used, such as removing the baby from both mothers had they not been able to reach an agreement on their own. However, this kind of solution is beyond the scope of our study for two reasons. First, as already noted, we do not allow an outside party – even a Solomon – to influence how a fair division will be made; we specify only a procedure and show what strategies the players can choose to ensure a minimal outcome. Second, we insist on some division of the goods or issues in question which, unlike the Solomonic solution (had it been used), must afford the players some satisfaction in terms of criteria we shall specify.[5]

1.2 Divide-and-choose and applications

As a first step toward specifying criteria for determining satisfaction, consider the well-known procedure for dividing a cake between two people, whom we will call Bob and Carol: Bob cuts the cake, and Carol chooses one of the two pieces. We assume that the cake is *divisible*, so it can be cut at any point without destroying its value. Yet it need not be *homogeneous*, or the same throughout; rather, we suppose it to be *heterogeneous*, wherein the flavors are mixed together but not stirred well.

Thus, rather than thinking of a layer cake, in which the flavors are completely separate, or a cake in which the flavors have been completely mixed and have the same consistency throughout, imagine a cake wherein there are uneven swirls of, say, chocolate and vanilla. Even though a piece may be 3/8 cholocate and 5/8 vanilla, there is no way to separate the chocolate from the vanilla by cutting the cake.

Switching the imagery and introducing preferences, suppose Carol likes the left side of a cake because it is thicker with frosting than the right side. Suppose Bob, who is on a diet, has the opposite preference but still would like the cherry in the middle. And both players may like the nuts, but they are scattered, with concentrations on both the left and right sides but not many in the middle. Patently, these preferences for the different toppings make this cake not only heterogeneous but also difficult for the players to divide in a way that will satisfy both.[6]

[5] A division of goods or issues, however, need not be into mutually exclusive parts; there may be sharing, as we will illustrate in section 1.3.

[6] However, we know of a case (from *Mother Goose*) in which complete satisfaction was achieved:

"One divides, the other chooses" is a *procedure* or *algorithm*, which gives the rules for a division, but it is not a *solution*, which also entails a description of how to apply these rules to achieve some stated level of satisfaction. The solution to this problem, however, is well known: Bob divides the cake into two pieces, between which he is indifferent; and Carol chooses what she considers to be the larger piece.[7] It is easy to see that this solution possesses two properties:

1 Bob can guarantee himself what he perceives to be at least half the cake (or, half the value of the cake; see note 7), and Carol can guarantee herself what she perceives to be at least half the cake. A fair-division procedure with this property is said to be *proportional*.[8]

2 Neither Bob nor Carol thinks that the other player received a larger piece of cake than he or she did. A fair-division procedure with this property is said to be *envy-free*.

Envy-freeness and proportionality are equivalent when there are only two players – that is, the existence of one property implies the existence of the other. Thus, if each player receives what he or she thinks is at least half the cake (proportionality), neither thinks the other player received more (envy-freeness). Conversely, if neither player envies the other player (envy-freeness), each must think he or she received at least half the cake (proportionality).

There is no such equivalence when there are three or more players. For example, if each of three players thinks he or she received at least 1/3 of the cake, it still may be the case that one player thinks another received a larger piece (say, 1/2), so proportionality does not imply envy-freeness. On the other hand, if none of the players envies another, each must believe that he or she received at least 1/3 of the cake.[9] In general, envy-freeness is the stronger notion of fairness: whenever it exists, so does proportionality, but not

> Jack Sprat could eat no fat,
> His wife could eat no lean;
> So 'twixt them both they cleared the cloth,
> And licked the platter clean.

[7] We will use expressions like "piece she considers largest" and "piece she values most" interchangeably, although the intuitions behind these expressions are quite different. The "size interpretation" is more useful in thinking about objects like cake, whereas the "value interpretation" is more useful in thinking about objects in which size is not the measure of worth. Note that to say that each player received at least half the value requires only an ordinal comparison – each player believes his or her piece is at least as large or as valuable as the other player's – but not a specification of how much larger or more valuable.

[8] If there are n players, *proportionality* means that each thinks he or she received at least $1/n$ of the cake. We will generalize this concept to players with different entitlements later.

[9] That is, if Bob, Carol, and, say, Ted divide the cake among themselves, and Bob thinks he received less than 1/3, then he must think that Carol and Ted are sharing more than 2/3. Thus, he will think that at least one of them (Carol, for instance) received more than 1/3, so he will envy her. It follows that when Bob (and the other players) are not envious, then each must believe that he or she received at least 1/3.

vice-versa. However, because one cuts, the other chooses is a two-person procedure that is proportional, it is also envy-free.

The origins of divide-and-choose go back to antiquity. The first mention we have discovered is in Hesiod's *Theogeny*, written some 2,800 years ago. The Greek gods, Prometheus and Zeus, had to divide a portion of meat. Prometheus began by placing the meat into two piles, and Zeus selected one.[10]

Divide-and-choose has important applications today. For example, the 1982 Convention of the Law of the Sea, which went into effect on November 16, 1994, with 159 signatories (including the United States), specifies that whenever a developed country wants to mine a portion of the seabed, that country must propose a division of the portion into two tracts. An international mining company called the Enterprise, funded by the developed countries but representing the interests of the developing countries through the International Seabed Authority, plays the role of the other party, choosing the tract it prefers; the developed country receives the other tract. In this manner, parts of the seabed are preserved for commercial development by the developing countries, which – in the absence of Enterprise – could not otherwise afford to mine the seabed. (Mining, however, probably will not begin in earnest for another decade.)

For divide-and-choose and other fair-division procedures we shall describe, the crucial assumption is that individual preferences are *weakly additive* in the following sense (Gale, 1993): if A is larger than B, and C is larger than D, and there is no overlap between pieces A and C, then A together with C is larger than B together with D. Although more realistic under the size interpretation than the value interpretation (see note 7), additivity is still sensible in many "value" situations, such as seabed mining, that we will encounter.[11]

As an illustration of the applicability of divide-and-choose – and some of the later procedures we will present – in a familiar context, consider the following estate problem (which, although hypothetical, is based on fact). The background is the following:

Ted and Alice owned and operated a motel on a lake in Vermont. Both were of retirement age when Ted died, leaving everything to Alice. Alice sold the motel and moved into a condominium in Florida, where she enjoyed a degree of financial security from the proceeds of the sale of the motel, retirement savings, and social security.

[10] Lowry (1987, pp. 126–31) discusses at some length the problems that plagued this transaction.

[11] Connections between weak additivity as used by Gale (1993) and the existence of finitely additive utility functions are given in Barbanel and Taylor (1995). We will henceforth assume that preferences are represented by finitely additive measures defined on the cake (or other goods) being divided. While "allowing arbitrary preferences would be unproductive" (Berliant, Dunz, and Thomson, 1992, p. 2), this is not to say that fair-division procedures based on substantial weakenings of finite additivity would be unwelcome.

Ted, however, had also left behind several items for which Alice had no use. Selling these was a possibility, but Alice neither needed the cash nor wanted the headache of dealing with a number of different retailers – or holding a garage sale, which would be fairly bizarre given the particular items available.

She therefore decided to let her two grown children, Bob and Carol, divide these items between themselves. Bob and his wife lived near the old homestead in Vermont. Carol and her family lived in a suburb of Boston, where she was an attorney. The items that Bob and Carol had to divide were as follows:

One 12-foot aluminum row boat
One 3-horse-power outboard motor
One piano in fairly good shape
One small personal computer
One hunting rifle
One box of tools (for carpentry and mechanics)
One 1953 Ford tractor with backhoe
One relatively old pick-up truck
Two mopeds (small motorized bikes).

Divide-and-choose provides a starting point for obtaining a satisfactory solution to Bob and Carol's division problem.[12] A coin is flipped and Bob wins, thereby becoming the divider. His task is to sort the ten items into two lists, knowing that Carol will have the choice of whichever list she prefers.

Conflicting motives may underlie how Bob chooses to do this, including the extent to which he knows his sister's preferences and the extent to which he is willing to exploit this knowledge for his own gain (or, more benevolently, for their mutual gain). However, if he is simply trying to put together two packages that he is indifferent between, he might split the items up as follows:[13]

Package 1: boat, computer, one moped, tractor

Package 2: motor, piano, one moped, truck, tools, rifle.

We assume that Bob is acting conservatively, guaranteeing himself a portion of the estate that he considers to be as valuable as the portion that Carol gets. Carol, of course, has the same guarantee since she gets to choose the "half" she wants. Thus, there should be no hard feelings, at least none caused by envy, but

[12] As we will see, there are several potential problems with the use of divide-and-choose in this situation. Later schemes - although slightly more complicated – will turn out to be vastly superior.

[13] Later in the chapter we will return to the more realistic question of how Bob and Carol would proceed, given that they certainly have some knowledge of each other's preferences. Exactly how is Carol going to get that 1953 Ford tractor to her suburb of Boston, and what is she going to do with it once she gets it there? Does Bob really have any use for either the computer or the piano? Later solutions to this fair-division problem will involve being more specific about the values that each player attaches to each of the ten items; our present treatment is consistent with these later values.

we should not forget Benjamin Franklin's admonition: "If you want to know the true character of a person, divide an inheritance with him."

Thus, Carol might resent the fact that Bob did not use what he knows about her preferences to put together two packages that each would definitely prefer to what the other received. Similarly, Bob might resent the fact that he is getting what he thinks is exactly half the value, whereas Carol is getting (it turns out) what she thinks is somewhat more than half.[14] For the moment, however, let us leave Bob and Carol to their pianos and tractors and turn to the legislative arena and a variant of divide-and-choose.

1.3 Filter-and-choose and applications

The first application of divide-and-choose to politics of which we are aware was proposed by the English political theorist, James Harrington (1611–77), in his book, *The Commonwealth of Oceana* (1656).[15] Dubbed "Harrington's law" by Goodwin (1992, p. 94), it was offered in the context of Harrington's analysis of a utopian polity, Oceana, intended as a model for England (Blitzer, 1960, p. 32; Pocock, 1992, p. xvii).

According to Harrington, the Senate of Knights of Oceana would, after debate, propose legislation, and the House of Deputies would, without debate, vote on it, putting the Senate in the position of Bob and the House in the position of Carol.[16] Thereby, Harrington argued, the vested interests of neither chamber could become dominant; it was a central tenet of his theory of egalitarian justice that also provided for the regular rotation of officials in office through annual elections to triennial offices, whereby one-third of the representatives would be elected each year (Goodwin, 1992, pp. 131, 197–8; Smith, 1914, p. 47).

Harrington was adamant that the House in Oceana (1,050 members), which would be more than three times larger than the Senate (300 members), should not debate legislation, even to the point of making this the only crime punishable by death in the commonwealth. According to Blitzer (1960, p. 241),

[14] This will become clear when we say more about how each player values each item. Carol, as we will see, will choose package 2.

[15] Immanuel Kant and Jean-Jacques Rousseau also proposed this idea (Goodwin, 1992, p. 94), though somewhat later and probably independently.

[16] The analogy is not exact, because when the Senate makes a particular proposal, its members – or at least a majority of its members – are presumably not indifferent to its passage, as is the cutter between which of the two pieces of cake the chooser selects. Moreover, the Senate, in making its proposal, may have some information on what will be acceptable to the House, which is a matter that we take up in section 1.4. Nonetheless, Harrington used the "homey little simile" (Smith, 1914, p. 51) of cake-cutting to justify his proposal. Others over the centuries, such as Count Petr Aleksandrovich Tolstoi (1761–1844), have repeatedly invoked it: "If the cake [a reference to the Ottoman Empire] could not be saved, it must be fairly divided" (quoted in Gulick, 1955, p. 72).

Harrington viewed the writing of laws by the more aristocratic Senate as

> the most important part of the governmental process . . . a task for experts . . . [requiring] native intelligence, knowledge of political theory and practice, and above all calm and dispassionate discussion.

Distinguishing between dividing and choosing, it is the Senate that has the

> indispensable function of debating, or dividing. [It] should necessarily be performed by the wisest and most virtuous members of the community, meeting together in a body small enough to permit calm and fruitful discussion (Blitzer, 1960, p. 241).

By contrast, the House is

> made up simply of representatives of the people. . . . An individual member is not expected to be particularly intelligent; rather, his duty is simply to reflect the wishes of his constituents (Blitzer, 1960, pp. 241–2).

Thus, Harrington proposed a bicameral legislature, similar in some respects to what the United States, forty-nine states (only Nebraska has a unicameral legislature), and most other democratic countries have today. However, his republic "rested upon a relation of equality between persons [the two houses] who were unequal in their capacities" (Pocock, 1977, p. 66).

In the US Congress, there is no distinction between the divider and the chooser. Although revenue legislation must be initiated in the House, each body can debate and pass its own version of a bill. If this happens, differences between the two versions are reconciled in a so-called conference committee, comprising members from both houses.

Once a compromise bill is agreed to by the conference committee, it cannot be amended when it goes back to the House and the Senate for action. A deal has been struck – that is, a division has been made, with some provisions of each bill included and some excluded in the conference committee compromise. Unlike in Oceana, however, the compromise bill can be debated in each body of Congress before a vote is taken.

Because the vote is either up or down, Harrington's law survives in emasculated form in Congress, with the conference committee the divider and each house a chooser. A bill passes if both houses assent to the compromise bill, and the president signs this bill into law.

A crucial difference between divide-and-choose in cake cutting and its application in the legislative process is that the divider does not get one piece and the chooser another. Instead, they both *share* the same piece – the bill that is eventually passed, or the status quo if it is not.[17]

[17] In the parlance of economics, the bill is a *public good*, from which each player can benefit without detracting from the benefit of the other player (Olson, 1965; Hardin, 1982). The "good" for some players, however, may turn out to be a "bad" for others, as we will see later.

Divide-and-choose in the legislative arena might be better characterized as *filter-and-choose*. Under the latter procedure, imagine that, instead of a single cake, a filterer must choose a subset of different pastries from a larger set – by filtering out what he or she considers the best – in such a way that the chooser will also prefer this portion. Too much of the set will give the filterer indigestion, whereas too little will leave him or her hungry, or otherwise unsatisfied by the selection, and likewise for the chooser.

But instead of the chooser's selecting between the subset and its complement (i.e., the other pastries), the chooser must select between the subset favored by the filterer and no pastries. This choice is analogous to passing new legislation (getting the subset, constructed by the filterer) or voting it down (keeping the status quo). Congresswoman Lynn Martin poignantly described the agony of this choice when she explained why she would support the 1986 tax-reform bill that emerged from a conference committee:

> I found, worried as I am about what this bill does, I am even more worried about the current code. The choice today is not between this bill and a perfect bill; the choice today is between this bill and the death of tax reform.[18]

As with the divider under divide-and-choose, the filterer under filter-and-choose – that is, the player who selects the subset – can ensure a proportional and envy-free division. This person does so by selecting a subset that has the same value as receiving no pastries, which may require dividing up one of the pastries to create this kind of equality.

Besides conference committees, another example of filter-and-choose in Congress is the "closed rule," which precludes amendments to bills on the floor of the House. This rule is applied to bills that might otherwise be strewn with amendments, such as appropriation and revenue bills.

A House committee that reports out a bill under this rule plays the role of the filterer, with the entire House the chooser. Since the choice is between this bill and no bill, one may properly think of the committee members as sifting through provisions that, they hope, will garner the approval of the entire House.

In the Senate, by contrast, no such restriction on amendments from the floor can be imposed. In fact, quite the opposite: a single senator can conduct a filibuster that precludes a bill from coming up for a vote, thereby preventing the Senate from even making a choice. However, a filibuster can be broken by a cloture vote, whose passage requires at least 3/5 of the Senate, so individual senators do not have an absolute right to block a bill.[19]

[18] Quoted in the *New York Times*, September 26, 1986, p. D7.

[19] Once debate is cut off, it may seem inconsistent that only a simple majority of senators is required for passage of the bill. The cloture rule's justification – that a larger majority should be required to halt the expression of "free speech" than pass a bill – is dubious if those members who vote for cloture are precisely those members who support the bill. When this is the case,

The power of a conference committee, and a House committee reporting out a bill under a closed rule, would appear to be greater than that of a regular committee, which might see its bill drastically altered by amendments from the floor. Remember, however, that if the (unamendable) bills emerging from the former committees fail to gain majority support on the floor, nothing can be done to save them, whereas a slight change in an (amendable) bill can sometimes make the difference between its passage and its failure.

In sum, the lack of control that a regular committee has over the contents of a bill, once it reaches the floor, must be weighed against the fact that an unamendable bill – emerging from a conference committee or under a closed rule in the House – cannot be fine-tuned, so to speak, in a way that might make it acceptable to a majority unwilling to support the unamended bill. On balance, however, having the ability to dictate a choice, without amendment, probably gives a committee greater power than not having this ability.

In the case of conference committees, to be sure, having the exclusive power to filter, and then propose, is somewhat illusory. These committees, after all, are not exactly free-wheeling entities; they must reach agreement within the strictures set by their respective houses, based on the bills each house passed.

By comparison, a House committee reporting out a bill under a closed rule has, ostensibly, more power, because its members – not one of the two houses – wrote the bill. But this fact may not prevent the House as a whole from balking at its committee's recommendation, especially if it is based on a closely divided committee vote, rendering the committee's power far from absolute.

There is still another incarnation of filter-and-choose in Congress. In impeachment proceedings of federal officials, including the president, the House assumes the role of the filterer: it is the body that draws up charges of impeachment, which require a 2/3 majority to bring about impeachment. If an official is impeached, which is equivalent to an indictment in a criminal court, the Senate then serves as the jury and can convict the official by a 3/4 vote, which makes it effectively the chooser in filter-and-choose.

We have traced the proposed use of divide-and-choose in governmental institutions back to the seventeenth century. But, as we indicated, divide-and-choose was known in ancient Greece more than two millennia before the appearance of Harrington's book. Empirical manifestations of its use in the Law of the Sea treaty, and in Congress today as filter-and-choose, indicate that it is not just played out in cake-cutting exercises on dining room tables.

We next show that the divider may exploit divide-and-choose if he or she possesses information about the chooser's preference. This vulnerability of the

one vote is tantamount to the other, so the 3/5 cloture requirement simply poses a greater hurdle for passage, vitiating its freedom-of-expression justification.

procedure, however, does not undermine its properties of proportionality and envy-freeness – or those of its variant, filter-and-choose.

1.4 The role of information

To exploit divide-and-choose, Bob must know Carol's preferences with respect to two things – (1) different portions of the cake, and (2) the value of obtaining a larger piece for herself compared with the value of denying him a larger piece. If Carol cares only about what *she* receives, Bob can, instead of dividing the cake equally for himself, cut it so as to give himself as large a piece as possible while still making Carol almost indifferent. We say "almost," because Bob needs to cut the cake so that Carol slightly prefers the piece he does *not* want and would, therefore, choose it for herself. Thereby Bob will end up with a larger piece for himself than if he simply created two pieces he considered equal.

On the other hand, if Bob thinks that Carol might see through his exploitative strategy and prefer to spite him for choosing it, he faces a dilemma. Equal division by Bob guarantees him a piece he considers to be 1/2 – it does not matter what Carol chooses. By comparison, creating near equality for Carol could give Bob more than 1/2, but at the risk of not even getting 1/2 if Carol chooses his (Bob's) larger piece out of spite. (To be sure, Carol would hurt herself doing so – in terms of the kind of cake she receives – but this may be a good long-run strategy if this kind of situation is likely to recur and Carol wants to deter future exploitation by Bob.) Thus, even knowing Carol's preferences for the cake does not ensure Bob of a piece he considers to be strictly larger than 1/2, given that Carol may resent being exploited and instead prefer to spite Bob rather than get a (slightly) larger piece for herself.

Note, by the way, that even to know she is being exploited, Carol must know Bob's preferences (i.e., whether Bob's cut reflects his true preferences or not). Only then can she act out of spite if she wants to do so.[20] Thus, even the two-person problem is not without some interesting twists, depending on the information available to the players.

To illustrate some specifics of exploitative strategies, let us return to the estate example (section 1.2), wherein Bob and Carol are trying to divide up

[20] Of course, Carol may prefer to act out of generosity, even knowing that Bob has been deliberately exploitative in making his cut. By the same token, Bob may prefer to cut the cake to as to give Carol what he thinks she will like. We by no means rule out such benevolent preferences but find malevolence more interesting to analyze, because it raises questions about how to protect against it. In games in which the players know each other's preferences, and spite is not a factor, the advantage of the divider is discussed in van Damme (1991, pp. 133–6) and Young (1994, pp. 137–45), which also include a discussion of alternative procedures to neutralize this advantage (see note 24).

the ten items left by their father and for which their mother has no use. Recall that Bob is doing the dividing, and he knows that Carol has little use for the boat, motor, truck, and tractor.[21]

By the same token, Bob has little use for the piano and computer, which would be almost impossible for him to sell. Both value the mopeds about the same. Carol, however, wants the tools more than Bob does, whereas Bob wants the rifle more than Carol does, though the rifle is an item that Carol can easily sell.

Given this information, and Bob's willingness to exploit Carol for his own gain, Bob could separate the ten items into the following two packages:

Package 1: boat, motor, one moped, rifle, tractor, truck

Package 2: piano, computer, tools, one moped.

It turns out – as will be seen when we quantify Bob's and Carol's preferences in section 5.5 – that Carol will, in fact, choose package 2, because she thinks it represents slightly more than half the value of the estate. This choice, of course, assumes that Carol is not spiteful about what Bob is doing in making this particular division.[22]

If Carol had won the coin toss that decided who was to be the divider, she could have taken advantage of her knowledge of Bob's preferences in the same manner that he exploited his knowledge of hers. Thus, she might construct the packages as follows:

Package 1: boat, motor, rifle, tractor

Package 2: piano, computer, tools, truck, two mopeds.

Assuming that Bob has no desire to spite Carol, he would choose package 1 since he thinks it represents slightly more than half the value of the estate. Notice that whether the divider (assume it is Bob) is being conservative or exploitative, he is trying to make the division 50–50 – but in two different ways. The difference is that:

[21] We suppose that the only value Carol attaches to these is the money she can realize from their sale. Moreover, the effort of selling them, which is liable to be considerable, is a cost she must consider. Because we assume that the values of the different items, including their resale value, are additive (see section 1.2), they can in a rough sense be summed. This assumption fails to the extent that one item (e.g., the boat) is less valuable without another item (e.g., the truck in which to transport it). In chapters 4 and 5, where we analyze different point-allocation procedures and apply one to real-life division problems, we shall say more about such complementarities and how to take them into account in fair division.

[22] If the value to Carol of spiting Bob exceeds her loss from choosing the pile he would prefer (and she would not prefer), then both siblings will be worse off – based on the values to each of the items alone – if Carol is spiteful. But once spite is factored into the equation, Carol will derive greater satisfaction from forsaking the more valuable items in package 2. (Formal modeling of this situation would require a context, familiar to economists, wherein preference relations and utility functions are defined on allocations of the entire cake rather than on the individual portions that each player receives, as we do throughout this book.)

(1) if he is being conservative, he makes the division exactly 50–50 in his own estimation;
(2) if he is being exploitative, he makes the division almost 50–50 in Carol's estimation.

Again, we say "almost" in case (2), because the divider wishes the chooser to select the divider's less preferred pile. Hence, the divider will endeavor to partition the items so that the chooser will slightly prefer one pile to the other, with the pile that the chooser does *not* select containing as large a fraction of the estate – in the eyes of the divider – as he or she can arrange.

Neither of these divisions will be "equitable" in the sense of trying to equalize the fraction of the estate that each perceives he or she is receiving. Thus, although each will receive at least half the estate – in their own estimation – and thus will not want to trade with the other, Bob will think he received only 50 percent of the estate while Carol will think she received, say, 80 percent in case (1), whereas there will be a reversal of roles in case (2). Correcting this aspect of the one-sidedness of divide-and-choose will be an issue we address with the point-allocation procedures in chapters 4 and 5.[23]

To sum up for divide-and-choose, dividers are disadvantaged unless they know the chooser's preferences, in which case they can capitalize on this knowledge. With or without this knowledge, dividers create exact or approximate indifference, but for different players: exact indifference for themselves, if they are in the dark about the chooser's preferences; approximate indifference for the chooser, if they know his or her preferences.[24]

In the filter-and-choose examples from Congress we gave in section 1.3, it is reasonable to suppose that members of committees, before reporting out a bill that can be voted only up or down by one or both houses, think carefully

[23] An informal attempt to deal with this and related problems of divide-and-choose is given in Singer (1962); a more formal attempt is given in Crawford (1977).

[24] Generally speaking, a divider will have some, if not complete, knowledge of a chooser's preferences, enabling him or her to exploit the chooser by his division (unless the chooser is spiteful). Young (1994, pp. 136–45) describes several schemes that neutralize the divider's advantage, some of which implement such classic game-theoretic bargaining solutions as those of Nash (1950), Raiffa (1953), and Rubinstein (1982); Moulin (1984) gives a scheme for implementing the Kalai-Smorodinsky (1975) solution, which is also analyzed in van Damme (1991, pp. 156–9). One rather theoretically compelling scheme, in our view, is that of Crawford (1979), in which two players bid to be the divider, which has been generalized to n players (Demange, 1984). But this scheme, like others, relies on using the fallback position of equal division of all items if the players are not able to agree on a mutually better division. While equal division makes sense for the two mopeds in our estate example, dividing a truck in two makes little sense, although its division based on a time-sharing arrangement might well be feasible. Difficulties like this tend to make procedures relying on the divisibility of all the items less desirable than procedures that do not. We will revisit this issue in our discussion of the point-allocation procedures in chapter 4, where other difficulties of using the bidding schemes of Crawford (1979) and Demange (1984) (e.g., their assumption of common knowledge) will be discussed.

about what is likely to be acceptable to each house. Conference committee members, especially, do not want to propose a bill that will likely go down to defeat in one or the other house. In fact, they work assiduously to balance provisions passed by each house so that the compromise bill will be acceptable to the memberships of both houses.

It is not just committees, though, that play the role of filterer in the federal government. Presidents initiate most major legislation, and they must be wary of proposing bills that are "dead on arrival." Normally they consult with their own party leaders in Congress, and sometimes opposition leaders, to draft bills that have a good chance of gaining majority support. But, of course, presidential initiatives are always subject to revision in congressional committees and – once the bills are reported out of committee – by each house, so presidents may have their roles as filterers greatly attenuated.

When it is the opposition party in Congress that initiates legislation, there is a reversal of roles. Now it is the administration that must respond by trying to accommodate itself to, or to contest, a proposal, which may well involve putting forward a counterproposal. In fact, there may be several bills that are introduced that reflect the various interests of different members of Congress.

Who, if anybody, is decisive in the filtering process is often unclear until one bill prevails in the end. Even final passage seldom shows up one player as critical, because most legislation is the product of many deals and compromises.

The distinction between filterer and chooser would appear to be more clear-cut when a bill is passed by both houses of Congress, and the president must decide whether or not to sign it into law. Because the president cannot amend it, he or she would seem very much in the role of chooser.

But this distinction is more apparent than real. Presidents influence the content of bills not only by marshaling their own forces in Congress to help shape the legislation but also by the threat of a veto should the bill either not contain provisions they deem essential or contain unacceptable features. In turn, Congress may respond as a filterer by trying to accommodate the president and writing in these provisions (if the veto threat is taken seriously), or by defying the president by trying to ensure that the bill is veto-proof – it will have the 2/3 support in both houses needed to override a presidential veto.

To review, the president, congressional committees, and Congress are both filterers and choosers at various times. Consequently, the issue is almost never one in which a filterer exploits knowledge about a chooser's preference, or a chooser decides whether to spite the filterer for this advantage. Rather, a president, Congress, and congressional committees move in and out as occupants of these roles: when a president proposes legislation (or floats a "trial balloon" to get an initial reaction before committing himself to it); when congressional committees write a bill; when a conference committee reconciles different versions of a bill after initial passage by both houses; when the compromise bill is sent back to each house; when, after final passage by both houses, the president signs the bill into law or vetoes it; and when Congress attempts to override a presidential veto.

This process is extended when new players are involved, such as the nine members of the Supreme Court if the constitutionality of the legislation is challenged. All fifty states become players in the ratification of constitutional amendments, which requires acceptance by 3/4 of the states once 2/3 majorities of both houses have proposed an amendment. And, finally, there may be strategic considerations quite removed from the legislative process, such as a president's vetoing a bill to enhance his reelection chances, even though he would prefer the bill to the status quo.

In conclusion, although certain congressional and constitutional procedures smack of filter-and-choose, it hardly ever occurs in the pristine form that Harrington proposed it for Oceana. True, a president or Congress – or the Supreme Court or the states – may be presented with a bill or a constitutional amendment that they must either accept or reject, but the process by which this bill or amendment was nurtured and shaped may well include their earlier intrusion. Thus, the separation of the roles of filterer and chooser is never airtight in the US federal system.

The lack of role separation, and the involvement of many players, are not necessarily detrimental to making better collective choices. The blurring of roles means that there is no dominant filterer, or gatekeeper, and the multiplicity of players means that several different points of view get expressed.

Players who anticipate and respond to each other at different stages are probably more conducive to producing balanced (though perhaps weaker) legislation. The process may also make for closer outcomes, especially if the players possess relatively complete information about each other's preferences. Then the filterer will defer to the chooser, but only insofar as necessary, to gain this player's acquiescence and not provoke a spiteful response.

The fact that much important and controversial legislation often passes Congress by only the slimmest of margins (e.g., one or two votes) is *prima facie* evidence that (1) there is reasonably good information about preferences in these situations and (2) the filterer takes this information into account in crafting bills. Although filterers may be advantaged in such situations, the process seems counterbalanced by the frequent reversals of role that the Constitution and legislative rules (e.g., that provide for conference committees) mandate, as we previously illustrated.

Harrington prohibited debate by the House so that its members would, as choosers, be handed well-conceived alternatives from the Senate that they could not adulterate. By acting only on the alternatives handed to them by the more informed and erudite Senate, Harrington had the quaint belief that House members – untainted by their ignorance and lack of intelligence – would make better decisions. Thereby Harrington could preserve democratic choice in the House, but one enlightened by the prior choices of the "rich and well born" (to use Alexander Hamilton's apt phrase) in the Senate.[25]

This attitude may strike one as anachronistic if not woefully elitist. In fact, however, Harrington's proposals were more democratic – in the populist sense of giving citizens some final, if indirect, say through their elected representatives in the House – than those advocated by most of his contemporaries. What he sought, at precisely the time of revolutionary change in England when old historical forms had been destroyed, was "a chance to construct new forms immune from the contingencies of history" (Pocock, 1992, p. xvii).[26]

Harrington believed, in his own unabashed account, that "if any man in England could show what a commonwealth was, it was myself" (Pocock, 1992, p. x). This may not be far from the truth; what Pocock (1992, p. xvi) most credits Harrington with is "a historical intelligence capable of synthesizing schemes of social change which set the *de facto* disorderliness of the English Civil Wars in a context of long-range and even universal historical processes."[27]

Like Harrington, the American founding fathers, who included Alexander Hamilton, also proposed a bicameral legislature. The functions of the two houses they proposed, however, were quite similar, which would make them better able to check each other as well as the president.

As these institutions have evolved, and new ones like political parties and congressional committees have formed, the roles of filterer and chooser have become increasingly mixed and diffuse, as we have seen. To some the process

[25] To be sure, Harrington wanted it to be an "aristocracy of excellence" (Pocock, 1977, p. 66), not just an aristocracy based on its members' place in society. But it is unclear how the capacities of this aristocracy would be recognized, selected, and accorded their proper standing. Although the possession of property (land, goods, or money) rather than birth qualified one as fit to serve in the Senate, this qualification imposed a substantial barrier on those not receiving an inheritance. They, as Harrington stressed, would have to acquire property through their own efforts (Pocock, 1977, p. 67).

[26] In the years just prior to the publication of *The Commonwealth of Oceana* in 1656, Charles I was executed (1649) and Oliver Cromwell became England's lord protector (1653). After Cromwell's death in 1658, a Republican commonwealth took control before the Stuart monarchy was returned to power in 1660, when Charles II reclaimed his father's throne.

[27] Views differ greatly, however, on Harrington's importance as a political theorist; see Downs (1977, pp. 13–15). But "in the sphere of practical politics," Cohen (1994, p. 128) asserts, "Harrington was ultimately more influential . . . than Hobbes – or for that matter, Vauban, Liebniz, Graunt, or Petty – since his doctrines were implemented in the following century, notably in the form of government adopted in the American Constitution." Cohen (1994, pp. 124–37) shows how Harrington, a great admirer of the biology of William Harvey, attempted to construct a "political anatomy," analogous to Harvey's anatomy and Harvey's analysis of circulation in the animal body.

has become incoherent, even chaotic, compared with Harrington's uncontaminated roles of Senate (divider) and House (chooser).

But with the spread of information (and misinformation) beyond the proverbial smoke-filled rooms, none of these institutions seems unduly advantaged today. If one institution benefits from being the filterer at one stage, other institutions assume its place later, so no dominant filterer or chooser is discernible.

We turn next to an extension of divide-and-choose, but one still applicable only to two players (we extend proportionality to more than two players in chapter 2). It suggests how the roles of divider and chooser may, in a sense, be made more symmetrical, which we illustrate with our Vermont estate example.

1.5 Austin's moving-knife procedure and applications

There is a "moving-knife" version of divide-and-choose that directly applies to cake cutting but which can be adapted to the division of discrete objects as well. In describing this moving-knife procedure – and others that will be developed later – we will follow Austin (1982) in picturing the cake as rectangular (see figure 1.1).

The procedure works as follows: A referee holds a knife at the left edge of the cake and slowly moves it across the cake so that, at every point along the horizontal edge, it remains parallel to its starting position at the left edge. At any time, either player can call "cut." When this occurs, the player who called cut receives the piece to the left of the knife, and the other player receives the piece to the right. A player who uses the obvious strategy of calling cut precisely when he or she thinks the knife splits the cake exactly in two will be guaranteed a piece he or she thinks is of size at least 1/2. Of course, the same issues arise with regard to the role of information of the two players, as we discussed for divide-and-choose in section 1.4.

This moving-knife version of divide-and choose is a special case of a more general procedure (to be discussed in section 2.4) introduced by Lester E. Dubins and Edwin H. Spanier in 1961. (We will give citations later.) They recast an older discrete algorithm due to Stefan Banach and Bronislaw Knaster, which we shall describe in section 2.3, as a continuous moving-knife procedure. It is the first such scheme that we know of in the literature.

Any discrete algorithm (like divide-and-choose) or moving-knife procedure (like the one just described) consists of both "rules" and "strategies." *Rules* incorporate those parts of the procedure that can be enforced by a referee, because the referee can tell whether or not a rule has been followed without knowing how the players value different parts of the cake. Thus, consider the following statement:

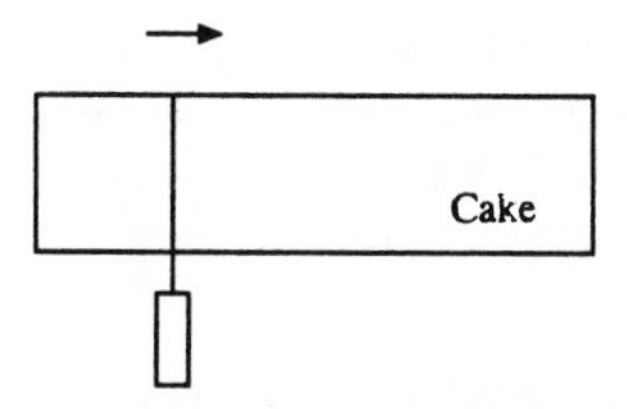

Figure 1.1 One moving knife

Bob cuts the cake into two pieces he considers to be of size 1/2.

This is not a rule, because it is not enforceable without knowing how Bob values different parts of the cake (i.e., his private information).

Strategies, on the other hand, are advice to players about the choices they *should* make, consistent with the rules, by using their private information about how much they value different parts of the cake.[28] Of course, as we saw in section 1.4, if players know something about the preferences of other players, they can use this information in an attempt to gain more than the minimum guarantee provided by the recommended strategies.

When presenting either a discrete procedure or a moving-knife procedure, we must give both the rules and the strategies. One way to do this, of course, is first to give the rules, which in the case of divide-and-choose are:

One of the players cuts the cake into two pieces in any way he or she chooses; the other then chooses either piece.

We may then describe the strategies that guarantee the players a proportional division – in this case, a piece of size at least 1/2 as each player values the cake:

The first player makes the division exactly 50–50 in his estimation; the second chooses a piece she considers at least tied for largest.

This is fine for divide-and-choose. For more complicated schemes, however, it is easier to understand a procedure if one is given the strategies at the same time as one is given the rules.

This is what we will do in the future. But to emphasize the distinction between rules and strategies in the case of Austin's procedure (to be described

[28] In game theory, a strategy is a complete plan that specifies the course of action a player will follow for every possible choice that other players might make. Players typically have many strategies from which to choose; here we single out one strategy that guarantees a player a particular outcome. Thus, our use of the term "strategy" is normative – a recommended course of action that enables a player to obtain something, regardless of the choices that other players make (as long as they follow the rules).

shortly), we will, after stating a rule, indicate in brackets the strategy that a player must follow to guarantee himself or herself a certain portion of the cake.

We begin by illustrating this distinction in the case of divide-and-choose. Its rules, and the strategies of its players that guarantee proportional (and envy-free) portions, can be described as follows:

Bob divides the cake into two pieces [that he considers to be of size 1/2], and Carol chooses one of the pieces [that she considers to be at least tied for largest].[29]

Although divide-and-choose, and its moving-knife counterpart, are the cornerstones for much of what we do in this book, they nevertheless have two fundamental deficiencies that, for reasons we shall discuss later, also manifest themselves in real-world applications:

1 The division they produce may not be *efficient* ("Pareto-optimal" in the language of economists): there may be another division that both players prefer to the one given by divide-and-choose, or that one player prefers and the other player considers as good.
2 The role of information or misinformation (as discussed in section 1.4) will almost assuredly leave one of the two players feeling that the other is more pleased with the outcome than he or she is.

As Rabin (1993) points out, mainstream economic theory has given far more attention to deficiency 1 (the lack of efficiency) than deficiency 2 (which involves what economists would call "interpersonal comparisons of utility" and on which we will, in chapter 4, base our notion of equitability). We will return to these properties many times throughout this book.

Our concern here, however, is with deficiency 2. Indeed, it is our contention that in many real-world situations, ranging from estate divisions to international treaty settlements, any allocation that leaves one player feeling not as pleased with the outcome as the other player will simply not be regarded as acceptable by the aggrieved player.[30] Thus in divide-and-choose, even if the role of divider is determined by the flip of a coin, which is presumably a fair procedure not favoring either player, one of the players may believe that the other has a definite advantage.[31]

It turns out that there is a way to modify the moving-knife version of divide-

[29] Because Bob's bracketed strategy (making the 50-50 division) maximizes the minimum that he can obtain whichever piece Carol chooses, his strategy is called a ***maximin strategy***.

[30] Sometimes both players may feel unhappy about a compromise settlement, but this does not mean – unless the settlement is inefficient – that both could have done better. The arbitrary manner in which the great powers have asserted control over countries, and redrawn borders, has certainly produced great inequities in the past (see section 4.6 for examples).

[31] As we pointed out in section 1.4, exactly which player this is – the divider or the chooser – will depend on what information each has about the other's preferences, including his or her inclination to be spiteful. The discussion in the remainder of this section concerns finding a procedure for leveling these roles, so that neither the divider nor the chooser is advantaged.

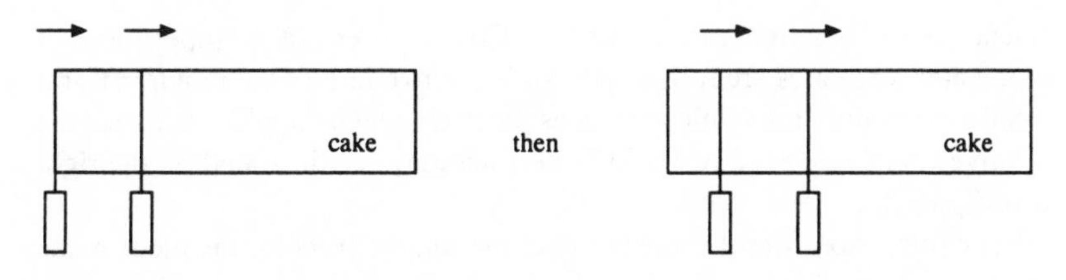

Figure 1.2 Two simultaneously moving knives

and-choose so as to recover some of the symmetry lost in both divide-and-choose and in its moving-knife analogue. This modification, due to A. K. Austin (1982), produces a division of a cake into two pieces in such a way that each player can guarantee the division is exactly a 50–50 split in his or her estimation.[32]

Thus, for example, if each player knows that the other will not risk getting strictly less than 1/2 (in that player's own estimation), one can divide the cake using Austin's procedure and then flip a coin to see who gets which piece (each valued at 1/2 by both players if they follow the prescribed strategies). Given our assumption that neither player will risk getting less than 1/2, both players have no choice but to make the division 50–50 in their own estimation.

Austin's procedure provides a way for the players always to achieve a 50–50 division simultaneously.[33] It begins with a knife moving slowly across the cake from the left edge toward the right edge, as in the Dubins–Spanier procedure, until one of the players – assume it is Bob – calls "stop" [which he does at the point when the piece so determined is of size exactly 1/2 in his estimation].[34] At this time, a second knife is placed at the left edge of the cake (see left side of figure 1.2). Bob then moves *both* knives across the cake in parallel fashion [in such a way that the piece between the two knives remains of size exactly 1/2 in Bob's estimation]. Hence, the physical distance between the two knives may vary.

We require that when the knife on the right arrives at the right-hand edge of the cake, the left knife lines up with the position that the first knife was in at the

[32] The description that follows is adapted from Brams, Taylor, and Zwicker (1994) with permission.

[33] In the mathematics literature, there are nonconstructive results that guarantee the existence of allocations that go well beyond what Austin's procedure provides. See, for example, Neyman (1946), Hobby and Rice (1965), Stromquist and Woodall (1985), and Alon (1987).

[34] Size is predicated on the knives' making vertical cuts. This is not to say that horizontal – or any of an infinite number of other planes through which the knives may cut – could not be better for both players (e.g., if Bob likes only the frosting, and Carol only the unfrosted cake, which only a horizontal cut could give them). The point is that our procedures provide guarantees for the players, whatever the specified plane is through which the knives exactly cut in half the cake.

moment when Bob first called "stop." (This requirement is superfluous if the indicated strategies are followed.) While the two knives are moving, Carol can call stop at any time [which she does precisely when the piece between the two knives is of size exactly 1/2 in her estimation], as illustrated on the right side of figure 1.2.

After calling stop, Carol chooses either the middle piece or the piece made up of the two sides flanking the middle piece. Notice that letting the player who *first* calls stop (Bob) choose the piece he wants – after the second player (Carol) calls stop – will not accomplish anything except to encourage both players, initially, to call stop as soon as possible. This is because the player who does so obtains what he or she thinks is the larger piece, whether this is the middle piece or the two pieces that flank it.

What guarantees that there will be a point where Carol thinks the piece between the knives is of size exactly 1/2? Observe that at the instant when the two knives start moving (left side of figure 1.2), Carol thinks that the piece between the knives is of size strictly less than 1/2 (otherwise she would have called stop earlier). At the point when the two knives stop moving on the right-hand side of the cake (assuming Carol does not call stop), the piece between the knives is the complement of what it was when the knives started moving.

Hence, Carol thinks this piece between the knives on the right-hand side is now of size strictly greater than 1/2. Assuming continuity in both the composition of the cake and the players' preferences for it, there must be a point where Carol thinks the size of the piece between the knives is exactly 1/2, making the division *even* in her eyes.[35]

Austin's procedure has an interesting property not arising under the other procedures we have considered so far: the strategy of waiting for the other player to call stop initially (Bob was the first to call stop in our example), regardless of how long it takes, is strictly better than calling stop when the size of the piece is 1/2 in a player's estimation (as suggested). For example, suppose you wait, and the other player calls stop at a point where you happen to think the size of the piece so determined is 3/4. Now, as soon as the two parallel knives start moving (in fact, the sooner the better), call stop. The result is that you will receive a piece only slightly smaller than 3/4 in your estimation (assuming you choose the middle piece), whereas you will think the other player is getting only about 1/4 from the two sides.

Of course, if your opponent also realizes the superiority of this strategy, then he or she will wait for you to call stop – at the same time you are waiting for him or her to call stop. Thus, the knife will travel from the left side of the cake

[35] In chapter 4 we will analyze a more general form of "equitability," under which both players believe they receive not just 1/2 but, instead, the same amount greater than 1/2.

all the way to the right side of the cake without either one of you ever calling stop.

How the cake is allocated in this situation is, in fact, a rule we did not specify, since we were only concerned with each player's having a strategy that would guarantee him or her a certain amount of cake, regardless of what the other player did. However, there are any number of ways to dissuade a player from employing this wait-for-your-opponent-to-call-stop strategy, such as flipping a coin at the end to see which player gets what piece, giving nobody any cake, and so on.

No generalization of Austin's procedure to ensure equality of pieces for $n > 2$ players is known. However, Austin himself noted that a simple extension of his procedure produces a single piece of cake that each of two players thinks is of size exactly $1/k$ for any $k \geq 2$.

This extension proceeds as follows: Bob first makes a sequence of $k - 1$ parallel marks on the cake [in such a way that he thinks the k pieces so determined are all of size $1/k$]. Now Carol cannot possibly think all k pieces are of size less than $1/k$, and she cannot possibly think all k pieces are of size greater than $1/k$. Thus, Carol thinks that either one of the pieces is of size exactly $1/k$ – in which case we are done – or we can assume, without loss of generality, that she thinks the first piece is of size less than $1/k$ and the second piece is of size greater than $1/k$.[36]

But now we can have Bob place two knives on the left and right edges of the first piece and move them as before [so as to keep the size of the piece between the two knives at exactly $1/k$], subject to the requirement that when the knife on the right arrives at the right edge of the two pieces, the left knife is at the position where the right knife was originally. The argument we gave earlier for $n = 2$ shows that, at some point, Carol will think the piece between the two knives is of size exactly $1/k$.

Austin's scheme can be further generalized to yield a partition of the cake into k pieces so that each of two players thinks all k pieces are of exactly size $1/k$. The proof of this generalization is by induction, but a simple case should convey the idea. For $k = 3$, we begin by obtaining, as before, a single piece of cake that both players think is of size 1/3. We now pretend that the rest of the cake is the entire cake; we apply what we did previously in the $k = 2$ case (i.e., divide the remaining 2/3 of the cake 50–50). This yields a division of the original cake into three pieces, each of size exactly 1/3 in the eyes of the two

[36] Why "without loss of generality"? Because some of the pieces are of size less than $1/k$ (we cannot have k pieces of size greater than $1/k$) and some of the pieces are of size greater than $1/k$ (we cannot have k pieces of size less than $1/k$), there must be some meeting of these different kinds of pieces whereby a strictly less piece is adjacent to a strictly greater piece. In the text, we simply assume that these are the first two pieces, and that the changeover goes from less than $1/k$ to greater than $1/k$ (instead of vice-versa).

players. The same reasoning can be used to show how the $k = 3$ result can now be used to obtain a division of the cake into four pieces, such that each player thinks each piece is exactly of size 1/4.

To illustrate how a version of this procedure might be applied in the real world, let us return to our ten-item estate last discussed in section 1.4, which we wish to divide between a brother (Bob) and sister (Carol) who value certain parts of it differently. We next present a discrete (i.e., noncontinuous) version of Austin's moving-knife procedure.[37]

The process begins with a referee (the executor of the will or an attorney) listing the items in some order. Of course, if one item is worth more to both players than everything else taken together, then there is no recourse but to sell such an item and replace it with the money so obtained. But this recourse will probably not be necessary in most situations.

As the executor runs his or her finger down the list – in a manner analogous to the knife moving across the cake – either Bob or Carol can call "stop" whenever he or she thinks the split is approximately 50–50. If the order of items is the one we gave for this example in section 1.2, it would turn out to be Carol who calls stop when the referee's finger moves just below the rifle:[38]

boat
motor
piano
computer
rifle

---

tools
tractor
truck
moped
moped

Because Bob did not yet call stop, we know he thinks the top of the list is worth less than half its total value. It is now up to Carol to start transferring items simultaneously from above the line to below the line, and from below the line to above the line, in such a way that she still thinks what is above the line is essentially the same value as what is below the line.[39] This is the analogue of moving two knives in such a way that the size of the piece between the two knives remains exactly 1/2.

[37] What we are calling the "discrete version" is similar in spirit to the "infinitesimal shavings" of a procedure due to Levmore and Cook (1981) that we discuss in section 6.4.

[38] Of course, if there are a number of "big-ticket" items in the middle, there may be no point when either Bob or Carol thinks the division is nearly equal.

[39] Again, there may be no swap of above-the-line and below-the-line items that preserves, in Carol's estimation, a nearly equal division except, trivially, an all-for-all swap.

For example, Carol might move the computer below the line, while moving one of the mopeds (which she thinks is of about the same value as the computer) above the line. At this point, Bob sees the division as pretty close to 50–50 also. Thus, they can now flip a coin to determine who gets which half and be assured that (1) the resolution is envy-free because neither desires the portion the other received, and (2) there will be no hard feelings caused by Bob's knowing that Carol is more pleased with the outcome than he is, or vice-versa. This resolution also protects both siblings from experiencing guilt created by their own – perhaps unintentional – exploitation of the other, making the resolution "guilt-free" as well.

We shall have more to say about settling the Vermont estate later. For now, however, we turn to the task of extending divide-and-choose and its moving-knife analogue to the case where there are more than two players.

2 Proportionality for $n > 2$: the divisible case

2.1 Introduction

Although a player like the president may propose legislation, almost invariably provisions are added to and subtracted from such a proposal, votes are taken, new changes made, and so on, as a bill wends its way through the legislative process. Likewise, the implementation of this legislation, once enacted, is also subject to significant modification through executive orders, court cases, citizen response, and the like.

Both divide-and-choose (section 1.2) and filter-and-choose (section 1.3) are strictly applicable only to two players. Although we suggested in section 1.4 that filter-and-choose is, effectively, played again and again in the legislative process, this two-person procedure provides an incomplete model if there are different players at each stage. Consequently, we ask whether there is an extension of divide-and-choose to more than two players.

The answer to this question, discovered by Hugo Steinhaus but discussed for the first time in Knaster (1946), extends divide-and-choose to three people. This result and others we shall describe in this chapter initiated the modern era of research on fair division, which focused on cake cutting but involved procedures for allocating indivisible goods as well (see chapter 3).

Hugo Steinhaus was a mathematician who, together with his colleagues, Bronislaw Knaster and Stefan Banach, began their research on fair-division procedures in Poland during World War II (Steinhaus, 1948). Their research – and that of dozens of others over the past half century – dealt with, among other things, two fundamental difficulties:

1 Allocation procedures that work for two or even three or four players often do not generalize easily to five or more players.
2 Allocation procedures that yield envy-free allocations are considerably harder to obtain than procedures that yield proportional allocations.[1]

[1] The concept of "envy-freeness," however, was not used by Steinhaus, Banach, or Knaster.

To set the stage for the later analysis of n-person envy-free procedures (chapter 7), we survey in this chapter n-person proportional procedures, beginning with Steinhaus' extension of divide-and-choose to the case of three players (Steinhaus, 1948). This procedure, which is called "lone-divider" and is described in section 2.2, was extended to $n > 3$ in Kuhn (1967).

Even before this extension was devised, Banach and Knaster (Steinhaus, 1948) discovered in the mid-1940s the n-person procedure we now call "last-diminisher" (Steinhaus, 1948), which is described in section 2.3. Dubins and Spanier (1961) were the first to propose the n-person moving-knife version of the last-diminisher procedure (section 2.4), and Fink (1964) the "lone-chooser" procedure (section 2.6), which was extended by Austin (1982) and Woodall (1986b) (section 2.7). Not only do these procedures fail to be envy-free, as we show throughout the chapter, but none is efficient. We illustrate this inefficiency with examples in section 2.8 and then a general argument that no algorithmic procedure applicable to heterogeneous goods can be both proportional and efficient.

Some of these n-person procedures, or the ideas underlying them, have probably been used informally, as we suggest may have been the case in the division of both Germany and Berlin into four zones after World War II (section 2.5). We also illustrate how the different procedures can be used to allocate goods when players have different entitlements (section 2.8). In addition, we suggest how some of the procedures can be adapted to the allocation of "bads" (rather than "goods"), like household chores, in which players want to be burdened with as few as possible.

2.2 The Steinhaus–Kuhn lone-divider procedure

The following description of Steinhaus' three-person procedure is based on the presentation in Kuhn (1967). If there are only three players, call a piece *acceptable* to a player if he or she thinks the piece is at least 1/3 of the cake.[2] Steinhaus' *lone-divider procedure* (in the case of three players) begins with

[2] It is possible to remove any mention of numbers from this and some (but not all) of our later procedures by adopting an axiomatic approach. For example, Steinhaus' lone-divider procedure for three players yields an allocation in which each player receives a piece he or she considers "acceptable" as long as one chooses a notion of "acceptability" that satisfies the following three axioms (COMAP, 1994, p. 403):

1. Each player is able to divide the cake into three parts so that any one of the pieces is acceptable.
2. Given any division of the cake into three parts, each player will find at least one of the three pieces acceptable.
3. Any two players who view a piece as unacceptable can obtain a fair share from a two-person, fair-division procedure over the remainder of the cake.

one player, chosen at random (say, Bob), who cuts the cake into three pieces that he considers to be all of size 1/3.[3]

Carol then indicates which of the three pieces she finds acceptable and which she does not, subject to the proviso (superfluous, if the strategy we indicate is followed) that she must approve of at least one. Ted then does the same. There are two mutually exclusive cases that can then arise:

Case 1. Either Carol or Ted (assume it is Carol) indicates that two or more of the pieces are acceptable (again, she thinks they are at least 1/3 of the cake).

In this case, Ted takes any one of the pieces that he considers acceptable. Carol then has available at least one of the pieces she considers acceptable, and Bob considers the remaining piece to be acceptable since he made the initial three-way division equal in his estimation. Choosing in the order Ted, Carol, and then Bob, therefore, will result in a proportional division.

Case 2. Both Carol and Ted indicate that at most one piece is acceptable (say, piece 1 for Carol and piece 2 for Ted).

In this case, there is a single piece (or perhaps two pieces) that Carol and Ted agree is not acceptable (piece 3).[4] This piece is given to Bob, while Carol and Ted reassemble and redivide what remains according to divide-and-choose. Notice that both Carol and Ted think they are reassembling more than 2/3 of the cake to divide between them. Hence, in case 2, we also have a division that provides each of the players with a piece that he or she considers acceptable.

We note two things about this procedure. First, because it is impossible for either Carol or Ted to think that all three pieces cut by Bob are unacceptable (i.e., of size less than 1/3), case 2 must obtain if case 1 does not – that is, the two situations we have identified are the only ones that can arise, making them exhaustive as well as mutually exclusive.

[3] This suggested equal-division strategy for Bob assumes that the three players will henceforth follow the rules. But there may, of course, be a problem if they do not, as illustrated by one of Aesop's fables, quoted at the opening of this book, which has given rise to the expression "the lion's share."

[4] If Carol and Ted consider only one piece acceptable, but it is the same piece (say, piece 1), then there are two pieces (piece 2 and piece 3) that both consider unacceptable. Either one, chosen randomly, could be given to Bob, who values them equally. Since the piece chosen is less than 1/3 for both Carol and Ted, the remainder must be more than 2/3 for them; hence, Carol and Ted can ensure themselves of more than 1/3 each with divide-and-choose. Naturally, if Carol considers piece 2 the smallest, and Ted considers piece 3 the smallest, there will be a conflict between them over which (unacceptable) piece to give to Bob. But the proportionality of the lone-divider solution does not depend on this selection, which, as we suggested, could be made randomly. Alternatively, one might have Carol and Ted first decide, on a random basis, who will be the divider. Because the divider is disadvantaged by divide-and-choose if there is no information about the chooser's preferences, then it might be argued that it should be the divider who decides which of the two unacceptable pieces (presumably, his or her smallest) will be given to Bob.

Second, while proportional, the lone-divider solution is not envy-free. In case 1, although Bob and Ted will envy no one, Carol will envy Ted if he took the larger of the two pieces that she considered acceptable. (If Carol considered all three pieces acceptable, then she would have valued them at 1/3 each and experienced no envy in this case.) In case 2, if the redivision by Carol and Ted is not exactly 50–50 in Bob's eyes, then he (Bob) will think that one of the players received more than 1/3 when he received exactly 1/3, causing him to be envious.

In sum, only Ted will assuredly be envy-free, whichever situation arises. Consequently, neither the divider nor one of the other two players has any guarantee of not experiencing envy under the lone-divider procedure.

Kuhn (1967), as we mentioned earlier, extended this procedure to any number of players. We will not describe his algorithm, which depends on a combinatorial theorem due to Frobenius and König (Kuhn, 1967), but we will present the $n = 4$ version in a way that is completely analogous to what we did for Steinhaus' original $n = 3$ version.[5]

The procedure begins with Bob's dividing the cake into four pieces which we will call A, B, C, and D, and which Bob considers to be equal. Each of the other three players now indicates which of the pieces he or she finds acceptable (in the sense now of thinking these pieces are the ones of size at least 1/4). We consider two main cases, each of which has a few subcases.

Case 1. At least one of the other three players (assume it is Carol) finds at least three of the four pieces acceptable.

Case 1.1. At least one of the other two players (assume it is Ted) finds two or more of the four pieces acceptable.

Let the players choose in the following order: Alice,Ted, Carol, Bob.

Case 1.2. Both Ted and Alice find exactly one of the four pieces acceptable.

Case 1.2 (i). The piece Ted finds acceptable is not the same piece that Alice finds acceptable.

Let the players choose in the following order: Alice, Ted, Carol, Bob (or Ted, Alice, Carol, Bob).

Case 1.2 (ii). Ted and Alice find only one piece acceptable, and it is the same piece.

There are at least two of the pieces that Carol finds acceptable that neither Ted nor Alice find acceptable. Hence, Ted and Alice think that these two pieces together make up less than half the cake, and so we can

[5] It is interesting to speculate on whether or not Steinhaus himself was aware of the $n = 4$ version of his result, given that it requires no additional ideas beyond those occurring in his $n = 3$ version. Nothing that we can find in the historical record substantiates speculation that he was aware.

give one of these pieces to Carol and the other to Bob. Ted and Alice can then reassemble the other two pieces (yielding more than half the cake in both their estimations) and split this between themselves using divide-and-choose.

Case 2. Each of the other three players finds at most two of the four pieces acceptable.

Case 2.1. At least one of the four pieces is acceptable to no one but Bob.

Give such a piece to Bob, and reassemble the other three pieces. The other three players think this yields more than three-fourths of the cake, so we can simply let them divide this among themselves using Steinhaus' $n = 3$ version of lone-divider.

Case 2.2. Each one of the four pieces is acceptable to Bob and at least one other player.

Since there are four pieces of cake and only three players besides Bob, there must be at least one player (assume it is Carol) who finds at least two pieces of cake acceptable (assume they are A and B). Since we are in case 2, Carol finds at most two pieces of cake acceptable. Thus, we know that Carol finds A and B acceptable, but C and D unacceptable. Since every piece is acceptable to someone besides Bob, we know that Ted or Alice finds piece C acceptable. Assume it is Ted.

Case 2.2 (i). Alice finds piece D acceptable.

Give Alice piece D, Ted piece C, Carol piece B, and Bob piece A.

Case 2.2 (ii). Alice does not find piece D acceptable.

Then Ted must find piece D acceptable. Hence, we can let the players choose in the following order: Alice,Ted, Carol, Bob, subject to the proviso that Alice does not choose piece D.

This completes the description of the $n = 4$ version of the Steinhaus–Kuhn lone-divider method.

In her senior thesis at Union College, Carolyn Custer (1994) includes a much simpler extension of Steinhaus' technique to handle the $n = 4$ case as well as a similar treatment for the $n = 5$ case. To illustrate the new ideas involved, we will present Custer's procedure for $n = 4$.

As before, we begin by having Bob divide the cake into four pieces (A, B, C, and D) that he considers to be equal and, hence, all of which are acceptable to him. We now consider three cases:

Case 1. There is at least one piece (assume it is A) that is acceptable only to Bob.

In this case, we give A to Bob. Pieces B, C, and D are reassembled, yielding a single piece that Carol, Ted, and Alice all think is of size at least 3/4. Hence, these three people can now use Steinhaus' $n = 3$ procedure, and each will obtain a piece that he or she thinks is of size at least (1/3)(3/4) = 1/4, as desired.

Case 2. There is at least one piece (assume it is A) that is acceptable only to Bob and one other player (assume it is Carol).

In this case, we give A to Carol and we let Bob, Ted, and Alice use the $n = 3$ procedure on the piece obtained by reassembling B, C, and D. This works because Bob, Ted, and Alice also think that B, C, and D together constitute at least 3/4 (Bob thinks it is exactly 3/4) of the cake.

Case 3. Otherwise.

In this case, each of the four pieces is acceptable to at least two players in addition to Bob. If we visualize the names of the players being placed below the pieces that they find acceptable, then we have at least eight boxes to fill in with three names:

A	B	C	D
Bob	Bob	Bob	Bob
□	□	□	□
□	□	□	□

Notice that at least one person (assume it is Carol) must find acceptable at least three pieces, lest the number of boxes filled in be at most 2 + 2 + 2 = 6. But even if Carol finds acceptable all four pieces, either Ted or Alice (assume it is Ted) must approve of at least two pieces lest the number of boxes filled in be at most 4 + 1 + 1 = 6. Thus, a proportional allocation can now be achieved by letting the players choose in the order Alice, Ted, Carol, and Bob.

2.3 The Banach–Knaster last-diminisher procedure

Steinhaus, apparently unable to make his solution work for $n > 3$ (see note 5), raised the question of such an extension with his Polish students, Stefan Banach and Bronislaw Knaster. The answer came in the form of a procedure, now known as *last-diminisher*, which was discovered by Banach and Knaster but was first described in Steinhaus (1948, 1949). Its elegance is well captured in Steinhaus' (1948, p. 102) description:

The partners being ranged A, B, C, . . . , N, A cuts from the cake an arbitrary part. B has now the right, but is not obliged, to diminish the slice cut off. Whatever he does, C has the right (without obligation) to diminish still the already diminished (or not diminished) slice, and so on up to N. The rule obliges the "last-diminisher" to take as his part the slice he was the last to touch. This partner thus disposed of, the remaining

$n - 1$ persons start the same game with the remainder of the cake. After the number of participants has been reduced to two, they apply the classical [divide-and-choose] rule for halving the remainder.

A strategy for each player that guarantees a proportional division is for the cutter to cut off a piece that he or she considers to be exactly $1/n$. When the piece reaches other players, they must decide whether it is more than $1/n$. If so, they diminish it to $1/n$ and then pass it on; if not, they pass it on undiminished, assured that there will be another piece that they or other players cut that will be at least $1/n$.[6] We assume that the parts of the slice that are cut off in the diminishing process are recombined with the original cake.

The last-diminisher procedure is not envy-free. Although all players can ensure themselves of at least $1/n$ by not diminishing a piece unless they view it as greater than $1/n$, all except the last two players may envy those who receive pieces later. More specifically, because the earlier players have exited the game (in section 7.2 we will change the rules to allow them to stay in), they cannot prevent a piece that they regard as greater than $1/n$ from going to one of the remaining players.

The last two players will experience no envy, because they regard no earlier pieces as greater than $1/n$. Consequently, for them the last two pieces must constitute at least $2/n$ of the cake; with divide-and-choose, each can ensure himself or herself of at least half of this remainder and so will envy nobody else.

2.4 The Dubins–Spanier moving-knife procedure

The moving-knife version of divide-and-choose for two persons that we presented in section 1.5 is a special case of the following elegant version of the Banach–Knaster last-diminisher procedure.[7] Its presentation by Lester E. Dubins and Edwin H. Spanier (Dubins and Spanier, 1961) was, as far as we know, the first example of a moving-knife procedure to appear in the literature.[8] The procedure, which we illustrate for three people, goes as follows.

A referee holds a knife at the left edge of the cake and slowly moves it across the cake so that it remains parallel to its starting position. At any time, any one of the three players (Bob, Carol, or Ted) can call "cut." When this occurs, the player who called cut receives the piece to the left of the knife and exits the game. The knife now continues moving until a second player calls cut.This

[6] A simple flowchart illustrating this procedure for three players is given in Bennett *et al.* (1987).

[7] This section, and sections 2.6 and 2.7, are adapted from Brams, Taylor, and Zwicker (1995b) with permission.

[8] But an earlier "pouring procedure" (the good to be divided is lemonade, which is homogeneous) has definite similarities (Davis, 1955, p. 30, exercise 11).

second player receives the second piece cut, and the third player gets the remainder. If either two or three players call cut at the same time, the cut piece is given to one of the callers at random.

One difference between this procedure and last-diminisher is that under last-diminisher the players do not have to make continuous choices – as the knife slowly moves across the cake – of whether or not to call "cut." Instead, they decide, for each piece when it reaches them, whether it is more than $1/n$; if so, they diminish it to $1/n$ (in their eyes) and then pass it on rather than, as under moving knife, wait for the knife to reach the $1/n$ point.

Assume, provisionally, in the moving-knife procedure that each player employs the strategy of calling cut anytime it yields that player a piece of size exactly 1/3. Assume, also, that Bob is the first player to call cut, and Carol the second player. Then Bob thinks he received exactly 1/3, and Carol thinks she received exactly 1/3. But Ted will think that both received less than 1/3, because otherwise he would have called cut himself. Thus, he thinks he received more than 1/3 and, consequently, more than either Bob or Carol, so he feels no envy.

However, both Bob and Carol might envy Ted. In particular, Bob will definitely envy either Carol or Ted if he thinks Carol did not make an exactly even split of what he (Bob) viewed as the remaining 2/3 of the cake after he made his initial cut. Although Carol will never envy Bob because she chose not to call cut first, she will definitely envy Ted because she allowed herself only 1/3 when she considered more than 2/3 of the cake to remain.

But Carol can easily amend her strategy to rectify this problem. Instead of calling cut when she thinks she gets exactly 1/3, she can call cut – after Bob has taken his piece – when she thinks the remainder is evenly split between her and Ted. Patently, she will not envy Ted if she is the first to call cut because of the 50–50 division of the remainder; should Ted call cut first, then Carol will think that the last piece, which she gets, is actually the larger.

Unfortunately for Bob (or any player who calls cut first), if he alters his strategy to call cut when he thinks he gets some amount more than 1/3, he risks ending up with less than his proportional share of 1/3, thereby also making him envious of at least one of the other players. At a minimum, the strategy of calling cut at 1/3 guarantees the first cutter his proportional amount, even if it does not ensure envy-freeness.

Carol and Ted will also receive proportional shares, which will be more than 1/3 if there were no ties when Bob called cut. In addition, they will not envy each other or Bob using the amended strategy.

It is worth reiterating that the purpose of a solution is to guarantee some kind of minimal outcome – in this case a proportional share. As Austin (1982, p. 212) put it, one wishes to "ensure justice for an honest person even when there is collusion by the other people." But even if there is no collusion, this guarantee also works against the simple greediness of players who may seek more than their $1/n$ proportional shares.

For more than three players, the first $n-2$ players can ensure a proportional share for themselves by calling cut at $1/n$, and the last two players can ensure that they experience no envy by calling cut when they think the remainder is exactly halved. Because a player does not know beforehand where in the process he or she will end up, however, this strategy does not guarantee him or her freedom from envy.

The guarantee of proportionality when there are more than two players, nonetheless, is no mean achievement. However, there are other ingenious procedures that ensure proportionality, and some of these have a more readily apparent real-world interpretation than that of a moving knife. Like lone-divider (section 2.2) and last-diminisher (section 2.3), some of these procedures require of the n players only discrete choices, not the continuous choices that the moving-knife procedure of this section demands.

2.5 Applications of last-diminisher and its moving-knife version

Like divide-and-choose and its variant, filter-and-choose, the ideas of fair division underlying the last-diminisher procedure and its moving-knife version, if not the specific details, can be found in some present-day institutions. Consider, for example, how a piece of land might be divided among three or more players.[9]

A common constraint on land division is that the parcels that players receive must be contiguous, which rules out apportionments in which a player receives two disconnected pieces. Another common constraint is that the borders separating different parcels must be straight lines. (This is true of the boundaries of most of the forty-eight contiguous states in the continental United States.) Given these constraints, a division could be carried out by having a knife sweep across the land – even if the land is odd-shaped or its border is irregular, as might be the case if the land included a jagged coastline – and having the players call cut when they think they receive $1/n$ of the total.

Although nobody, literally, would use the moving-knife procedure in this way, we suggest that players might propose divisions as if there were a moving knife. For example, a player who especially liked the land along a

[9] For more mathematical approaches to land-division problems, see Hill (1983), Beck (1987), Webb (1990), and Legut, Potters, and Tijs (1994). The empirical literature on border disputes and the drawing of international boundaries includes Boggs (1940), Chisholm and Smith (1990), Rumley and Minghi (1991), Goertz and Diehl (1992), Finnie (1992), Lustick (1993), and Gardner (1993). Samuelson (1985) discusses both theoretical and empirical issues of dividing coastal waters; problems in the international sharing of water resources are discussed in Gleick (1993) and Lowi (1993a, 1993b).

coastline might be the first to call cut as the moving knife, starting from the coastline, sweeps inland.

But a knife sweep starting at the coastline and moving inland is only one of an infinite number of ways to cut. Furthermore, it could yield noncontiguous portions if the coast includes deep bays. By contrast, a knife sweep down a coastline might work better, giving everyone a piece of the beach.

Sheathing the moving knife for a moment, assume a player proposes a parcel of land along the coastline for himself. If this player's proposal was acceptable to the other players, another might say that she would like the adjacent land so-and-so distance from where the first player's ends, as if a knife had continued moving and she were the next to call cut. Again if there were no objections, proposals like this might continue, as if the knife were cutting farther and farther inland each time, until the land was completely apportioned.

Of course, there may be objections, and these could be handled by the last-diminisher procedure. Assume, for example, that the first player, in the eyes of another player, proposed more than a proportional share. The objector could respond by saying that he or she would settle for only part of the coastline, or all the coastline but less area inland. Still another player might reduce his or her demand and say that an even smaller parcel would be acceptable, and so on, until nobody had any further objections – that is, until nobody proposed a smaller piece, at or near the coastline, that he or she would accept.

When this point is reached, the last-diminisher takes this (diminished) parcel. The division of the remaining land then continues among the $n-1$ other players.

This kind of thinking, if not the last-diminisher procedure itself, is often applied informally to the allocation of different kinds of divisible goods, including land. For example, it is reflected in the negotiations among the allies in 1944 to partition Germany into four zones, which were to be under their jurisdiction (Smith, 1963). Several plans were put forward, and adjustments made over many months, delaying a formal agreement until 1945.[10]

[10] Decisions were quicker in the post-war allocation of other territories in Europe. As Winston Churchill put it in a 1956 interview, "We agreed on the Balkans. I said he [Joseph Stalin] could have Rumania and Bulgaria; he said we could have Greece (of course, only in our sphere, you know). He signed a slip of paper. And he never broke his word" (Sulzberger, 1970, p. 304). Lest this sound callous, we note that as early as 1941 Churchill had set forth some principles of land division in a note to Stalin: "Territorial frontiers will have to be settled in accordance with the wishes of the people who live there and on general ethnographical lines, and secondly that these units, when established, must be free to choose their own form of government and system of life, so long as they do not interfere with the similar rights of neighboring peoples" (Miner, 1988, p. 147). Both these statements of Churchill are excerpted in Gardner (1993, pp. 265 and 94, respectively). Not only is the contrast between the first statement and the more noble sentiment expressed in the second statement striking, but other remarks of Churchill and his foreign minister, Anthony Eden, also clash with this sentiment (see section 4.6).

We shall return to this case in section 7.5, when we discuss multistage envy-free procedures that, it seems, are closer to what the allies actually used after they wrestled, in particular, with the problem of what to do with Berlin. The fact that Berlin ended up wholly inside the Soviet zone, but was itself partitioned into four zones administered by the different allies, illustrates that not all land divisions are contiguous.

Thus, a procedure like that of moving-knife, which tends to preserve contiguity, could not have been used in the division of Germany and Berlin. On the other hand, the last-diminisher procedure (section 2.3) more readily allows for noncontinguity. As an example, the part diminished could be the land that connects two formerly connected parts.

The last-diminisher procedure would appear to have some of the earmarks of a Dutch auction. In such an auction, the price for an item begins so high that nobody is willing to bid for it. The price then falls steadily until one player stops the fall by making a bid at the price reached, thereby winning the item at this price.

In effect, the price itself is continually being diminished until, at some point, it becomes acceptable to one player. But this price is also the top price that any player would be willing to pay, because otherwise another player would have stopped the fall earlier.

Despite the resemblance of a Dutch auction to the last-diminisher procedure, the correspondence is only superficial. It is not clear, for example, what receiving a piece of size $1/n$ means when a person is bidding on goods in an auction.

An English auction, by contrast, is one in which the price is bid up until no player makes a higher bid than the last one made. In the fair-division context, it might be dubbed *last enhancer*.

We shall have more to say about auctions in chapter 9, where we will propose a new two-stage procedure. Suffice it to say here that the divisibility of land makes it amenable, at least in principle, to the application of the proportional procedures so far discussed.

2.6 The Fink lone-chooser procedure

In 1964, A. M. Fink (Fink, 1964) discovered a discrete algorithm for achieving a proportional allocation among n people that is quite different from either the Steinhaus–Kuhn lone-divider (section 2.2) or Banach–Knaster last-diminisher (section 2.3) procedures. It has the desirable property that the procedure can be initiated without knowing exactly how many players are (or will be) involved. Again, we illustrate it in detail for the case $n = 3$, with the extension to $n = 4$ (or more) being quite easy (as it was for last-diminisher, but decidedly not so for lone-divider).

Assume that, in what follows, the players go in the order Bob, Carol, and then Ted:

1 Bob bisects the cake, creating two pieces of equal size, A and B, for himself.
2 Carol chooses what she considers to be the larger of the two pieces, giving Bob the other piece.
3 Bob and Carol trisect the pieces they now have. Ted then selects what he considers to be the largest one of the three portions of each of the trisected pieces, leaving the remainders to Bob and Carol.

Bob thinks he received exactly 1/3, since he is getting 2/3 of a piece he thinks is of size 1/2. Carol thinks she received at least 1/3, since she is getting 2/3 of a piece she thinks is of size at least 1/2. Ted thinks he received at least 1/3 since he is getting at least 1/3 of A and at least 1/3 of B, and hence at least 1/3 of the whole cake. Thus, the procedure is proportional.

Bob, however, may envy either Carol or Ted, because the piece that Carol trisects may, in Bob's eyes, be "uneven," giving either Carol or Ted (but not both) more than 1/3. (If they both got more than 1/3, he would have to get less, which is impossible since his cuts ensure himself of 1/3.) Similarly, Carol may be envious of Bob or Ted because of the unevenness problem, and Ted may be envious of Bob or Carol;[11] for example, if Ted thinks the piece Bob got (say, A) is 3/4 of the cake, his getting at least 1/3 of A and B (i.e., at least 1/3 of the entire cake) may not be as big a combined piece as the 1/2 that he thinks Bob got.

If $n = 4$, Bob, Carol, and Ted would quadrisect their three pieces at the next stage, giving Alice a choice of what she considers to be the largest one of the four portions of each of the three pieces. The algorithm continues in this fashion for any n, with n-section at the last stage.

2.7 Woodall's and Austin's extensions of Fink's procedure

In his 1948 paper, Hugo Steinhaus (Steinhaus, 1948, pp. 102–3) reported the following:

> It may be stated incidentally that if there are two (or more) partners with different estimations, there exists a division giving to everybody more than his due part (Knaster); this fact disproves the common opinion that differences in estimations make fair division difficult.

[11] No player, however, can envy all the other players under this or any other proportional procedure. That is, if I have at least $1/n$, then I cannot think everyone else has more than $1/n$. Notice also that if I think someone has more than $1/n$, then someone else must have strictly less than $1/n$ in my estimation, which generalizes our earlier observation (section 1.2) that if there are only two players, then proportionality (both players have at least 1/2) implies envy-freeness (both players have at least as large a piece as the other).

Steinhaus provided no proof of this remark,[12] nor did he give any indication of whether Knaster actually had a procedure for achieving this division or simply an existence result, without an algorithm for finding the division.

In 1986, Douglas Woodall provided an algorithm for achieving an allocation with Knaster's property (Woodall, 1986b).[13] It assumes that (1) there is a piece of cake that two players value differently, and (2) we know the fraction of value that each player attributes to that piece.

In order to give some of the flavor of Woodall's procedure (the details can be found in Woodall, 1986b), we make the stronger assumption that we have two pieces of cake such that Bob and Carol disagree as to which is larger. (Woodall shows how to obtain such pieces of cake from the weaker assumption that two players value one piece differently.) If we now have Bob and Carol apply divide-and-choose to the rest of the cake, then it is easy to see that we can obtain a partition of the entire cake into two pieces, A and B, such that Bob thinks A is larger than 1/2 while Carol thinks B is larger than 1/2. (A includes the piece that Bob thinks is larger, and B includes the piece that Carol thinks is larger.) This solves the problem for $n = 2$, because each player thinks he or she is getting more than a proportional share.

If we assume that a piece of cake is of no value to one player unless it is of no value to every player, then Woodall's (1986b) method of handling additional players can also be simplified. We do this by applying the rest of Fink's algorithm (section 2.6) at this point, obtaining an allocation in which each of the n players has a piece of cake he or she considers to be of size at least $1/n$, while Bob (and Carol, although we won't need this fact) has a piece he thinks is of size strictly larger than $1/n$.

But now Bob can split his piece into sufficiently many equal-size pieces so that the removal of any $n - 1$ of them will still leave him with a piece he considers to be of size strictly larger than $1/n$. (For example, if $n = 3$ and Bob's piece was of size 2/5, splitting this 2/5 into twenty pieces of 1/50 each means that two of these small pieces can be removed and still leave Bob with $2/5 - 2/50 = 9/25 > 1/3$ of the cake.) We now simply let each of the other $n - 1$ players choose one of these slivers (valued at 1/50 by Bob in our example) to

[12] Proofs were independently provided by Urbanik (1955) and Dubins and Spanier (1961). The cake-cutting literature does have antecedents with expressive names, including the "problem of the Nile" (Fisher, 1938; Dubins and Spanier, 1961) and the "ham sandwich theorem" (Stone and Tukey, 1942; Stromquist and Woodall, 1985).

[13] For a finer analysis of allocations with Knaster's property, see Elton, Hill, and Kertz (1986), Hill (1987b), and Legut (1988b). For related measure-theoretic considerations, see Legut (1987, 1988a), Legut and Wilczyński (1988), Hill (1987a, 1987b, 1993), Akin (1995), Barbanel (1995a, 1995b, 1995c), Barbanel and Zwicker (1994), and Barbanel and Taylor (1995).

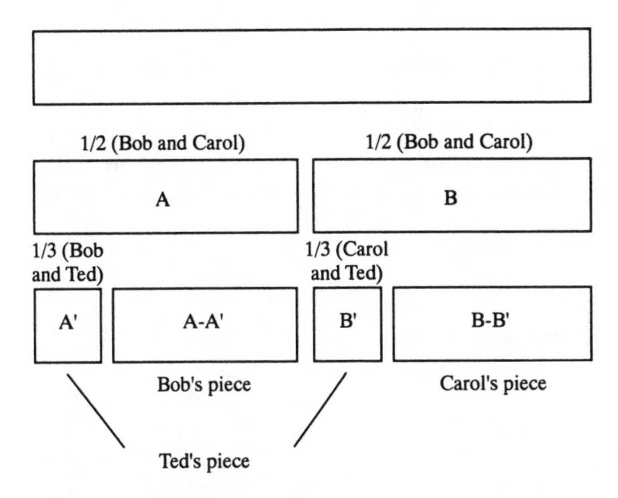

Figure 2.1 Austin's revised procedure

add to what they already have. Essentially, we have applied one of Woodall's ideas at the end of Fink's procedure, instead of repeating it at each stage of his procedure, as Woodall did.

Another revision of Fink's procedure, which does not give each player more than $1/n$ but exactly $1/n$, is due to Austin (1982). He noticed that by introducing into Fink's procedure his (Austin's) moving-knife procedure (section 1.5) for obtaining a single piece of cake that each of two players thinks is of size exactly $1/k$, one can obtain a moving-knife procedure that yields an allocation among n players such that each player thinks he or she receives a piece of size exactly $1/n$.

Austin's revised procedure proceeds as follows: Bob and Carol divide the cake into two pieces, A and B, that both agree are of size 1/2, based on Austin's earlier moving-knife procedure. Bob and Ted then cut a piece A' from A that they both think is exactly 1/3 of A, which is also possible under Austin's earlier procedure. Carol and Ted now do exactly the same thing to B to obtain B', giving the pieces shown in figure 2.1. Bob now receives A – A' (i.e., the result of removing A' from A, which he views as 2/3 of 1/2, or 1/3), Carol receives B – B' (2/3 x 1/2 = 1/3), and Ted receives A' and B' (1/3 of A plus 1/3 of B, and thus 1/3 of the entire cake). Thus, each player receives a piece that he or she thinks is of size exactly 1/3.

Now we use Fink's procedure. If a fourth person (Alice) comes along, each of the three earlier players simply gets together with Alice and cuts a small piece that he or she and Alice agree is 1/4 of the piece held by that player. Alice then gets the union of the three small pieces (1/4 of each yields 1/4 of the whole cake), whereas the three earlier players each receive the remainder (3/4 x 1/3 = 1/4).

Although every player thinks he or she received exactly $1/n$ under Austin's revision of Fink's procedure, it is not envy-free. Thus, for example, in the three-person case, Bob will envy either Carol or Ted unless he also thinks they cut a piece of size exactly 1/3 from B.

2.8 Efficiency, entitlements, and chores

1 Efficiency

An allocation is *efficient* if there is no other allocation that is strictly better for at least one player and as good for all the others.[14] In economics, such an allocation, which was first proposed by Vilfredo Pareto, a nineteenth-century Italian economist, is referred to as Pareto-optimal or Pareto-efficient.

We hasten to add that an efficient allocation need not be proportional, envy-free, or fair in any other sense (we shall define a concept of "equitability" in section 4.3). For example, if one player gets all the cake, or all the indivisible goods available, this allocation is efficient – except in the instance where this player places no value at all on a portion to which some other player attaches some value. Although a more uniform distribution may be better for all except this player, if this player suffers, even slightly, because he or she does not get everything, then the more uniform allocation is not one under which *everybody* (in particular, this player) does at least as well. Consequently, a very lopsided distribution in which one player gets all the good or goods being allocated, unfair as it might seem, can be efficient.

What the notion of "efficiency" conveys is that there is no better overall outcome for society. If some people improve their lot, others will be hurt, so no outcome that helps everybody – or at least some people, without hurting others – is possible when there is efficiency.

[14] If an allocation fails to be efficient, then there typically exists another allocation that is strictly better for *all* the players. To see this, assume that we have an allocation that is not efficient, and choose another allocation that is strictly better for one player (Bob) and no worse for the others. Now have Bob distribute *part* of his "surplus" among the other players. This yields an allocation that is strictly better for all the players as long as every player attaches positive value to any portion to which some player attaches positive value. (Our proviso on positive value is why we said "typically" in the first sentence of this footnote.) The usual solution of economists to inefficient allocations is to allow players to trade with each other to obtain better allocations; however, this is not a procedure but an informal adjustment mechanism. Still, it may be desirable to encourage trading in situations in which the fair-division procedure does not guarantee an efficient allocation.

We next show that none of the proportional procedures we have discussed either in this chapter or the previous chapter invariably yields an efficient outcome, except in special cases.[15] The assurance that they give each player of getting at least $1/n$ of the good or goods being allocated, therefore, provides no guarantee that there is not another assignment that gives some players more without hurting the rest. After illustrating this result for the various procedures, we shall indicate why, in fact, no procedure of the kind we have discussed has any hope of being both proportional and envy-free.

1 Divide-and-choose. A cake is 1/2 vanilla and 1/2 chocolate. Bob likes only chocolate, and Carol likes only vanilla. Assume Bob is the cutter (it could be either player). To ensure himself of 1/2 of his preferred flavor, Bob must divide the cake so that each half contains an equal portion of chocolate. But, regardless of which piece Carol chooses, the allocation wherein Bob receives all the chocolate and Carol receives all the vanilla is no worse for Carol as well as being strictly better for Bob.[16]

Of course, Bob's knowledge of Carol's preferences could save him from the unfortunate circumstance of having to make a 50–50 split of the chocolate. But we make no such assumption about his having this information. Consequently, Bob's strategy that guarantees him a proportional and envy-free share in this example yields an inefficient outcome.[17] (We shall extend this example to three players and three flavors when we discuss shortly the moving-knife procedure.)

2 Filter-and-choose. Here the metaphor is pastries, and suppose the filterer proposes a set that is slightly better for the chooser than receiving none (the status quo). (Recall that we supposed that too many of these sweets would give the chooser indigestion, just as a too watered-down bill in Congress would be worse than the status quo.) Indeed, if the filterer has complete information, he

[15] Every procedure is efficient when all the players have the same preferences for all parts of the cake. This follows from the fact that a different allocation obtained by adding or subtracting pieces adds or subtracts identical values, so there is no way some players can be made better off without hurting others.

[16] This allocation is not strictly better for Carol if Bob splits the cake into portions of (1) half chocolate and (2) half chocolate plus all the vanilla. With this split, Carol would choose (2), which is the same for her as getting all the vanilla alone.

[17] Consider what kinds of allocations are efficient when there are just two players, two flavors of cake, and both players always like more cake rather than less (unlike the example in the text), even of the flavor they less prefer. Then *all* divisions in which at least one player gets all of his or her preferred flavor are efficient. The reason is that if both players receive both preferred and nonpreferred flavors, then they can trade nonpreferred for preferred and both do better (unless their preferences are identical), making such an allocation inefficient. On the other hand, if at least one player (say, Bob) has all his preferred flavor, although he can be made better off with more of his nonpreferred flavor, the other player will be made worse off giving up this portion, making such an allocation efficient.

can select that alternative which is best for himself among those that would be just barely better than the status quo for the chooser.

If the chooser accepts, then this result is efficient. On the other hand, if the filterer has no information, or only incomplete information, he will not be able to make such a close calculation of acceptability. Even though, say, a watered-down bill passes, there might be another bill that both sides – presuming they have some common interests (e.g., in not running a huge deficit) – would prefer. In the no-information or low-information case, therefore, we may well get an inefficient result under filter-and-choose.

Also, if spite is a factor, the chooser may reject a slightly better alternative because she regards it as an insult and, in fact, prefers to be spiteful in response to such a paltry offer. This rejection outcome is inefficient compared with the filterer's being somewhat more generous and, as a result, having his offer accepted – at least if the latter outcome makes both players better off than the status quo (brought about by the chooser's rejection of the paltry offer).

To illustrate the inefficiency of filter-and-choose without the spite factor, we can use the 1/2 chocolate–1/2 vanilla cake just discussed but now assume that, for the filterer, a portion of vanilla makes a negative contribution that is equal in magnitude to the positive contribution made by the same-size portion of chocolate. Given these preferences, the filterer must propose a portion of the cake that consists of an equal amount of chocolate and vanilla, assuming he wants to guarantee that he will not wind up with negative value. However, if the chooser has the same preferences as the filterer, the allocation resulting from this proposal is worse for both than the obvious division that has the two sharing the chocolate portion of the cake – assuming the filterer can entirely filter out the vanilla from the allocation.[18]

3 Moving-knife. Imagine a rectangular cake consisting of a piece of chocolate flanked by vanilla strips on each side. Suppose Bob likes only chocolate and Carol likes only vanilla. Clearly, the only efficient allocation is to give Bob all the chocolate and Carol all the vanilla. However, the moving knife – no matter what angle it starts at – cannot produce this allocation.

4 Lone-divider. Suppose we have a cake that is 1/3 chocolate, 1/3 vanilla, and 1/3 strawberry. Assume Bob likes only chocolate, Carol likes only vanilla, and Ted likes only strawberry. It is obvious that the only efficient allocation is the one that gives each player all of the flavor that he or she likes. However, under the lone-divider procedure, the player going first (assume it is Bob) must divide the cake into three pieces he is indifferent among if he wants to

[18] To be sure, if the vanilla can be filtered out, the filterer can presumably discard the vanilla after the division has been made. But if our players are children who, their parents tell them, must eat everything on their plates, then this discard strategy will not work.

guarantee himself 1/3 of the total value. Thus, each of these three pieces must contain 1/3 of the chocolate, and so we do not obtain the unique efficient allocation.

5 *Last–diminisher*. The same cake and three players as discussed under the lone-divider procedure show that the last-diminisher procedure is inefficient. The strategy of the first player (assume this person is Bob) is to cut off a piece that he considers to be exactly 1/3 of the total value. Thus, he will cut a piece consisting of 1/3 chocolate and maybe some vanilla and strawberry as well. If it contains less than 1/3 of both the vanilla and strawberry, then Bob will end up with this piece, showing that we again do not arrive at the unique efficient allocation. On the other hand, if it is trimmed by one of the other players to what he or she considers to be 1/3 of the cake's value, then it still cannot end up being one of the three pieces in the unique efficient allocation.

6 *Lone-chooser*. With the same cake and three players as just discussed under the lone-divider and last-diminisher procedures, the strategy of the first player (assume this person is Bob) is to cut the cake in two so that each portion contains 1/2 of the chocolate. Assume Carol goes next, so she will choose the portion with more vanilla. After Bob and Carol trisect their portions so that each smaller piece contains 1/3 of the chocolate and 1/3 of the vanilla, Ted will select one smaller piece from Bob's portion and one smaller piece from Carol's portion that contain the most strawberry. While it is theoretically possible for Ted to end up with all the strawberry, Bob and Carol cannot possibly receive all of the chocolate and vanilla, respectively, because Ted took some from each, so the unique efficient allocation is unattainable.

In summary, none of the procedures guarantees the players an outcome that cannot be improved on for at least some players without hurting the others. Thus, in addition to the proportional procedures not being envy-free, they are not efficient.

One reason for this lack of efficiency stems from the requirement that the players, when they choose strategies that guarantee themselves proportionality, must forego strategies that could yield more. Thus, for example, divide-and-choose, lone-divider, and lone-chooser require the dividers to equalize portions, even though they may prefer different parts from the chooser(s) and could, therefore, afford to keep more for themselves.

But the need to ensure receiving at least $1/n$ prevents their taking such a chance. A similar argument holds in the case of the moving-knife procedure, under which players call cut early to avoid the risk of losing out later. Likewise, the last-diminisher procedure precludes a player from proposing a disproportionately large piece initially, lest he or she lose this piece (only slightly diminished) to another player and end up with a disproportionately small piece as a consequence.

Although we have gone through a procedure-by-procedure analysis, there really is no need to check for efficiency in each case – they are all doomed. More generally, there is no fair-division scheme that is simultaneously (1) algorithmic, (2) proportional, and (3) efficient.

Any two of these properties can be obtained,[19] but it can be shown that simultaneously obtaining all three is impossible. Roughly speaking, the argument goes as follows: Suppose we have an algorithm that produces a proportional allocation for two people. Now consider what it does when these two people value all parts of the cake the same.

Obviously, the resulting division of the cake will give two pieces of size 1/2 in the common estimation of the players. Since we are restricting ourselves to an algorithm, only the players' estimates for a finite number of pieces of cake have any effect on the behavior of the algorithm.[20]

But now change the preferences of one of the two players, but only with respect to subsets of one of the pieces of cake chosen so that none of these subsets played any role in how the algorithm just worked. Hence, the algorithm will again yield a 50–50 split in the eyes of both players, but now their preferences for all parts of the cake will no longer be identical. Because Woodall's version of Fink's procedure (section 2.7) shows that there is an allocation in this situation that both players prefer to a 50–50 split, the algorithm is not efficient.

2 Entitlements and chores

Each of the fair-division procedures discussed in this chapter can be adapted to the situation wherein the players are entitled to different (rational) fractions of

[19] The last-diminisher method, for example, satisfies (1) and (2), whereas simply giving the entire cake to one player satisfies (1) and (3). A result guaranteeing the existence of an allocation satisfying (2) and (3) – and even envy-freeness – was proved in Weller (1985), with additional results given in Berliant, Dunz, and Thomson (1992) and Akin (1995). With the exception of Akin, who is a mathematician, the authors of these papers are all economists, whose work tends to focus on finding conditions for the existence of a solution rather than finding a procedure (i.e., algorithm) for obtaining it.

[20] The idea of an algorithmic fair-division procedure, although compelling at an intuitive level, has not been rigorously formalized here or elsewhere. Hence, our assertion in the text that "only the players' estimates for a finite number of pieces of the cake have any effect on the behavior of the algorithm" is not one that can be derived from first principles but, rather, one that must be accepted as a *thesis* – that is, an assertion connecting an intuitive notion with a proposed mathematical formalization that is, therefore, not subject to mathematical verification. Readers who question this thesis will, of course, also want to question the conclusion based on it. For other examples of nonexistence results in the fair-division literature, see Even and Paz (1984) and Robertson and Webb (n.d.b).

the cake.[21] The way this is done, which goes back to Steinhaus (1948), is quite easy. For example, suppose we have three players – Bob, Carol, and Ted – entitled to fractions of 1/12, 3/12, and 8/12 of the cake, respectively. (We can always express the fractions using a common denominator.) Then we simply pretend that we have twelve players, one of whom has Bob's preferences, three of whom have Carol's preferences (i.e., are all clones of each other), and eight of whom have Ted's preferences. After the procedure is run, Bob gets his piece, Carol gets the three pieces corresponding to her three clones, and Ted gets the eight pieces corresponding to his eight clones.

Science writer Martin Gardner (Gardner, 1978) was apparently the first person to interpret fair division in terms of dividing up a set of chores, like mowing the lawn or taking out the garbage, wherein each player is trying to minimize, instead of maximize, the portion he or she obtains. It turns out that each of the procedures given in this chapter can be adapted to this situation as well.

For example, in applying the last-diminisher procedure, a piece would be cut by a player and passed along, with each player increasing the piece until it equaled $1/n$, thereby ensuring that the remaining chores (or parts of chores) are not subsequently divided in such a way – say, equally – that he or she might be forced to take more than $1/n$. If a player thought a piece was of size at least $1/n$, he or she would not add to it, because this would mean that he or she might end up getting more than a proportional share. Similarly, the aim of a player using the lone-divider and lone-chooser procedures would be to make minimal rather than maximal choices at each stage so as to ensure that he or she does not obtain more than a proportional share.

In the case of the moving-knife procedure, the rules would need to be modified so that the player who calls cut would receive the piece to the right of the knife instead of the piece to the left. The procedure would then be repeated for the remaining players. A player's strategy would be to call cut when the piece to the right of the knife is of size exactly $1/n$. (For example, if there were four players remaining, a player would call cut when the knife had swept over 3/4 of the cake, leaving 1/4 to the right.) Anyone calling cut sooner would receive more chores, which is what a player would not want to happen. When only one person remains in the game, he or she would be given the remainder of the chores, which this person will believe is not more than $1/n$ since the $n - 1$ other players, in his or her estimation, already each received at least $1/n$ of the chores.

There is another modification we can make in the moving-knife procedure

[21] An algorithm handling entitlements that correspond to irrational numbers is given in Barbanel (1995a). Other work on entitlements can be found in Legut (1985), McAvaney, Robertson, and Webb (n.d.), and Robertson and Webb (n.d.a).

that gives a different proportional solution to the chores problem. Instead of awarding the first player who calls cut the piece to the right of the moving knife, we award the last player to call cut the piece to the left of the moving knife.

Consider how this procedure would work when no piece has yet been cut. The first $n - 1$ players will call cut when they believe the knife has swept over exactly $1/n$ of the chores. However, the nth player will not call cut when he or she thinks the knife has reached the $1/n$ point but rather immediately after the $(n - 1)$st person has called cut. Thereby all players except the nth player will believe that the first piece cut is more than a proportional share of the chores, whereas the nth player will believe this piece that he or she receives is no more than a proportional share. The procedure is then repeated, but with one less player in the game after each piece is cut until only one person remains, who is then given the remainder of the chores.

This procedure is proportional for reasons analogous to those given for the first modification. Both procedures are, in a sense, mirror images of the n-person moving-knife cake-division procedure. In the first modification, "right" of the knife is substituted for "left" for the piece given to a player; in the second modification, "last" is substituted for "first" for the player who receives the piece.

The first modification resembles a Dutch auction (section 2.5): one starts too high (too many chores) and then comes down to the level acceptable to the player with the highest tolerance for chores on the right. The second modification resembles an English auction: the players successively indicate their upper limits for chores on the left until the player with the highest limit receives this piece. As we will show in section 7.2, if we allow players to reenter the game, both of these procedures can be made approximately envy-free.

To conclude, all the n-person proportional procedures for divisible goods that were described in this chapter – whether the procedures are discrete (the Steinhaus–Kuhn lone-divider procedure, the Banach–Knaster last-diminisher procedure, the Fink lone-chooser procedure) or continuous (the Dubins–Spanier moving-knife procedure) – are inefficient. Moreover, all fail to be envy-free, except in the case of two players. Each of them can be modified, however, to handle either entitlements or chores.

But they achieve only partial success in applications involving goods that are not completely divisible. This was true of the Vermont estate, as discussed in section 1.5, and may also be true of land, as discussed in section 2.5, if it contains property, like a house, which cannot be easily divided. It is precisely this new context – in which indivisible goods must be divided – to which we turn in chapter 3.

3 Proportionality for $n > 2$: the indivisible case

3.1 Introduction

The proportional procedures described in chapter 2 are not generally applicable to real-world situations involving *indivisible goods*, whose value is destroyed if they are divided. They were, in fact, designed for the case where the good being divided, such as cake or land, is divisible – that is, a cut can be made at any point.

In this chapter, by contrast, we assume that there are k different indivisible goods, which cannot be partially allocated. The fair division problem is to assign each such good to one and only one player so that each player thinks he or she is getting at least $1/n$ of the total.

We focus on two procedures for producing a proportional division of indivisible goods.[1] The first is the "procedure of sealed bids" proposed by Knaster some fifty years ago (Steinhaus, 1948). This procedure seems quite practicable if the players all have an adequate reserve of money. The second procedure is the "method of markers" developed by William F. Lucas about twenty years ago (Lucas, 1994), which builds on ideas behind the last-diminisher procedure and applies them to indivisible goods. These procedures are discussed in sections 3.2 and 3.3.

In section 3.4, we illustrate how these procedures might also be used to allocate goods when players have different entitlements, as in the case of estate division. We also explain in section 3.4 the sense in which Knaster's procedure is efficient but not envy-free, whereas Lucas's procedure is neither efficient nor envy-free.

[1] For a randomized procedure that was actually used in estate division, see Pratt and Zeckhauser (1990), which is based on the notion of a competitive equilibrium in Hylland and Zeckhauser (1979); see also Young (1994, pp. 156–9). Other randomized solutions are analyzed in, among other places, Demko and Hill (1988), Broome (1991), Szaniawski (1991), Goodwin (1992), and Elster (1989, 1992). On the envy-free allocation of indivisible goods in which monetary compensations are possible, see Maskin (1987) and Tadenuma and Thomson (1993, 1995). Although most of this work is axiomatic, exceptions are Alkan, Demange, and Gale (1991) and Aragones (1995), in which some results are constructive in nature.

3.2 Knaster's procedure of sealed bids

As a way of allocating "fairly indivisible objects (houses, animals, engines, works of art, etc.)," as distinguished from "physically divisible objects . . . [such as] pieces of land" (Steinhaus, 1948, p. 103), Knaster proposed an ingenious kind of auction that we call *Knaster's procedure of sealed bids*, or simply *Knaster's procedure*. Like a conventional auction, players submit sealed bids for items, and the player who bids the highest is awarded the item. Unlike a conventional auction, however, some of the money that is bid for the items is also divided up among the players.

As an application of this procedure, consider the situation in which a mother wills that three items must be shared equally among her four children. Clearly, without being privy to any more details, we know that not all the children can receive items, because there are too few to go around. Instead, at least one will have to be compensated in another way; the obvious way is for him or her to receive money from the children who are more successful in their bidding. This requires that each child start with a bankroll, or have access to collateral against which he or she can borrow money, that enables each to make bids – and possibly compensate less successful bidders, who win either less valuable items or no items.

Let us modify the estate example by assuming that there are at least as many items as heirs, making it possible for each heir to receive at least one item. Specifically, consider the example given in Luce and Raiffa (1957, pp. 366–7) and Raiffa, (1982, pp. 290–1), in which there are four indivisible items and three heirs. Let the four items be A, B, C, and D, and assume they are to be shared equally among Bob, Carol, and Ted. Later we will consider the possibility of unequal shares, based on different entitlements.

Suppose that each heir attributes the monetary values, shown in table 3.1 under *valuation*, to the four items.[2] Notice that not only does each heir attribute a different value to each item, but the *total valuation* of each heir of the four items is not the same amount.

Thus, for example, Ted's total valuation is the greatest ($14,000). Bob is not far behind ($13,300), but Carol places significantly less value on the four items ($8,500). We assume that the monetary values of each heir are *additive*, so if Ted, for example, obtains both items B and C (worth $4,000 and $2,000 to him, respectively), his combined valuation for these two items is $6,000.

In fact, because Ted values B and C more than either of the other two heirs, Knaster's procedure awards him both items, as shown in table 3.1 for *items received*, when he bids $4,000 and $2,000, respectively. Similarly, Knaster's

[2] We assume that these valuations are honest – what a person would be willing to pay for them – and that there is no collusion among the heirs, but we will relax these assumptions later.

Table 3.1. *Knaster's procedure with 3 players and 4 items*

	Player		
	Bob	Carol	Ted
Valuation			
Item A	$10,000	$4,000	$7,000
Item B	2,000	1,000	4,000
Item C	500	1,500	2,000
Item D	800	2,000	1,000
Total valuation	13,300	8,500	14,000
Items received	A	D	B, C
Value received	10,000	2,000	6,000
Initial fare share	4,433	2,833	4,667
Difference (initial excess/deficit)	5,567	–833	1,333
Share of surplus	2,022	2,022	2,022
Adjusted fair share	6,455	4,855	6,689
Final settlement	A – 3,545	D + 2,855	B, C + 689

procedure awards item A to Bob and item D to Carol, because each of these players bids more for these items than either of the others. (If there were a tie for top bid, it could be broken randomly.) The monetary values of the items that each heir receives are shown as *value received* in table 3.1.

The next step in the procedure is to compute an *initial fair share* for each heir, which will be revised later. This is simply each player's total valuation of the four items divided by three (rounded to the nearest dollar in table 3.1), since the three heirs are assumed to have equal claims on the estate. Thus, for example, Bob's initial fair share is $13,300/3 ≅ $4,433.

The *difference* between the total value of the item(s) each heir receives and his or her initial fair share, which may be either positive (an *initial excess*) or negative (an *initial deficit*) is shown on the next line in table 3.1. Specifically:

- Bob receives item A, worth $10,000 to him; his initial fair share is $4,433, giving him an initial excess of $5,567;
- Carol receives item D, worth $2,000 to her; her initial fair share is $2,833, giving her an initial deficit of $833;
- Ted receives items B and C, worth $6,000 to him; his initial fair share is $4,667, giving him an initial excess of $1,333.

Under Knaster's procedure, the initial excesses (of Bob and Ted) and the initial deficit (of Carol) are summed algebraically, giving

$5,567 (Bob) – $833 (Carol) + $1,333 (Ted) = $6,067.

As long as the players differ in their valuations of the items – as is the case here – this sum will be positive (Kuhn, 1967, p. 34), so we call it the *surplus*.[3] Knaster's procedure assigns each player an equal share of this surplus, or $6,067/3 ≅ $2,022.

When these equal shares of the surplus are added to the initial fair shares of each heir, we obtain an *adjusted fair share*, which is also shown in table 3.1. Thus, for example, Bob's adjusted fair share is $4,433 + $2,022 = $6,455.

Notice that the adjusted fair share of each heir may be either greater or less than the value to each heir of the item or items received. Calculating the difference between the monetary value of the item(s) an heir receives and his or her adjusted fair share, we see that each heir obtains either a *final excess* or a *final deficit*:

- Bob receives a final excess of $10,000 – $6,455 = $3,545;
- Carol receives a final deficit of $2,000 – $4,855 = –$2,855;
- Ted receives a final deficit of $6,000 – $6,689 = –$689.

Because of his final excess, Bob must contribute $3,545 to the others, which we call a *positive side payment*, so his *final settlement* is shown in table 3.1 as A – $3,545. On the other hand, Carol and Ted, because of their final deficits, must receive contributions of $2,855 and $689, respectively, which we call *negative side payments*, making their final settlements D + $2,855 (Carol) and B, C + $689 (Ted).[4]

Observe that the sum of the final deficits (of Carol and Ted) equals the final excess (of Bob)

$2,855 + $689 = $3,544 ($3,545 after rounding).

Put another way, the negative side payments and the positive side payments sum to zero, so the market "clears" (i.e., there are no outstanding balances).

Also observe that the monetary values of the final settlements – that is, the adjusted fair shares – of Bob ($6,455), Carol ($4,855), and Ted ($6,689) give each, respectively, 49 percent, 57 percent, and 48 percent of their total

[3] If the players' valuations for all items are exactly the same, the surplus will be zero. Disagreement among the players, therefore, benefits them by creating a positive surplus, enabling each to obtain a more-than-proportional share of the estate, as we will illustrate shortly.

[4] Fink (1994) has suggested to us an alternative way of describing Knaster's procedure, which we illustrate next. Bob bid the most for item A ($10,000), so let him make up a pot comprising A and 2/3 [i.e., $(n-1)/n$, where n is the number of players] of his bid ($6,667). Now Bob withdraws A from the pot, and there is enough left over to give everyone else his or her "fair share" (i.e., 1/3 of what he or she bid): $4,000/3 ≅ $1,333 to Carol, and $7,000/3 ≅ $2,333 to Ted; this leaves (essentially) a $3,000 surplus. Divide the $3,000 surplus among the three bidders, so everybody gets $1,000. Repeating this calculation for each item leads to the final settlements shown in table 3.1. The advantage of this description of Knaster's procedure over the one in the text is that it can be applied independently to each item, although, as Fink (1994) explains, "in a real situation one would probably have a referee do all of them [the calculations] and only announce the overall result."

valuations, which are considerably more than their 33 percent proportional shares. Not only do the allocations to each heir satisfy the criterion of proportionality, but these allocations are – at least in this particular case – envy-free:

- Bob, who receives \$6,455 in monetary value, values Carol's allocation as \$800 + \$2,855 = \$3,655, and Ted's allocation as \$2,500 + \$689 = \$3,189;
- Carol, who receives \$4,855 in monetary value, values Bob's allocation as \$4,000 – \$3,545 = \$455, and Ted's allocation as \$2,500 + \$689 = \$3,189;
- Ted, who receives \$6,689 in monetary value, values Bob's allocation as \$7,000 – \$3,545 = \$3,455, and Carol's allocation as \$1,000 + \$689 = \$1,689.

Unfortunately, the greater value that the heirs attribute to their own final settlements, compared with that of the others, is peculiar to this example. In general, Knaster's procedure does not guarantee envy-freeness, as the following example demonstrates:

Assume that there is only one item to divide among the three players, and assume Bob values it most highly. Then it will go to him. If Carol values it next most highly, then the side payment she will receive from Bob will be greater than the side payment that Ted will receive. Therefore, Ted will envy Carol for getting more money.

Knaster's procedure, nevertheless, is always envy-free in two-person situations. It takes an especially simple form if there is only one indivisible item that must be divided between two players, with the player who receives the item compensating the other player for not receiving it. Raiffa (1982, pp. 292–7) illustrates this calculation with two players, whom we call Bob and Carol, who must decide who gets an old encyclopaedia, E – and at what cost to the recipient (reckoned in compensation to the other player). Bob values the encyclopaedia at \$40, Carol at \$100.

In table 3.2, we show the results of applying Knaster's procedure not only for these valuations but also for symbolic valuations, x and y, for Bob and Carol, where $x < y$. Note that the adjusted fair share for each player gives him or her half the valuation ($x/2$ or $y/2$) plus half the surplus of $(y - x)/2$ [i.e., $(y - x)/4$].

Thus in the numerical example, because Carol bid more for E than Bob did, she gets E but must pay Bob \$35 in compensation. This gives both players \$15 more than their initial fair shares (i.e., half their valuations) of \$20 (Bob) and \$50 (Carol).[5]

When there are only two players, and each gets at least half his or her

[5] When there are just two players and one item, the final settlement can be immediately calculated from the two bids (\$40 and \$100 in our example): the high bidder gets the item and pays the lower bidder half the average of the two bids as compensation (\$70/2 = \$35). Thus, each player is accorded an item or compensation according to his or her own expressed desire, plus equal parts of the surplus.

valuation under Knaster's procedure, neither player will envy the other player, whose share is viewed as necessarily less than half. In our numerical example, Bob receives 87.5 percent of his valuation (35/40), whereas Carol receives 65.0 percent of her valuation (65/100), so unlike the point-allocation procedures we will analyze in chapter 4, Knaster's procedure is not "equitable" – the players do not receive the same percentage increments above 50 percent. We will shortly investigate whether equitability can be ensured if the players under Knaster's procedure have the same total valuations for the item(s) to be divided.

A potential drawback of Knaster's procedure is that players can profitably misrepresent their preferences if they know the monetary values that the other player(s) attribute to the items. In the two-person example just described, Raiffa (1982, pp. 296–7) shows that for every additional $4 that Bob exaggerates his $40 valuation, he gets an additional $1 in the final settlement.[6] Thus, increasing his valuation by almost $60 (to ensure that it stays just under Carol's valuation of $100) adds almost $15 to Bob's final settlement, giving him almost $50 rather than just $35. By this misrepresentation, paradoxically, Bob would receive in compensation more than what he thinks the item is worth ($50 rather than $40) – quite a benefit of fair "division."

But such misrepresentation can be risky if Bob should not be sure of Carol's valuation. For example, if Bob bids $120 for E, thinking that Carol bid slightly more, then he would win E but have to give Carol a side payment of $55, suffering a loss of $15 (because 40 – 55 = –15). If both players misrepresent, they may both lose over what honest bids would give them. Raiffa (1982, p. 296) concludes that "although one can't really say that the Steinhaus scheme [really the Knaster procedure] encourages honest evaluations, in many situations it may be the pragmatic thing to do."

In applying Knaster's procedure, the main practical difficulty lies in its assumption that players have money at hand, or perhaps can borrow it, in order to make the side payments that arise. To be sure, if they have final deficits rather than final excesses, they do not have to spend their money, because they are the parties being compensated rather than the ones having to do the compensating. But, of course, there is no assurance beforehand that they, rather than other players, will be freed of the burden of making positive side payments.

In the case of the next proportional procedure that we shall discuss, we

[6] Another form of manipulation, collusion among players, may also be profitable. Thus in our earlier three-person example, if Bob and Carol know that Ted most values item B – and, more specifically, will bid $4,000 for it – then they can best exploit him by inflating their bids for this item from $2,000 (Bob) and $1,000 (Carol) to $3,999 each, ensuring themselves of greater compensation by losing by only $1 on this item.

Table 3.2. *Knaster's procedure with 2 players and 1 item*

	Bob (bid of 40)	Carol (bid of 100)	Bob (bid of x)	Carol (bid of y)
Valuation of E	40	100	x	y
Items received	–	E	–	E
Value received	0	100	0	y
Initial fair share	20	50	$x/2$	$y/2$
Initial excess		50		$y/2$
Initial deficit	20		$x/2$	
Share of surplus	15	15	$(y-x)/4$	$(y-x)/4$
Adjusted fair share	20 + 15	50 + 15	$x/2+(y-x)/4$	$y/2+(y-x)/4$
Final settlement	35	E – 35	$(x+y)/4$	E – $(x+y)/4$

assume that $k \geq n$ (there are at least as many goods as players, and generally many more), and that an allocation assigns at least one good to each player. If this were not the case, then a player receiving no good would not only be envious of the others but also think, because he receives nothing,, that his allocation is not even proportional. This did not arise in Knaster's procedure because monetary side payments were made.

3.3 Lucas' method of markers

In this section we again consider the question of how to divide a collection of indivisible objects among n players. Our starting point is the moving-knife procedure of Dubins and Spanier (section 2.4) and, more specifically, a version of this procedure that suggests itself when we arrange the items in a list – or, literally, lay them end to end along a line. In fact, this version of the moving-knife procedure is exactly what we did with Bob and Carol in section 1.5 when we had the referee move his or her finger down the list of ten objects until either Bob or Carol called stop.

One drawback of the moving-knife procedure lies in the extent to which a particular ordering of items favors some players over others. For example, if there are four players, then the first player to call cut gets what he or she thinks is (essentially) 1/4 of the total value of the items. With that player and those items gone, the remaining three players can employ the strategy of not calling stop until the knife has passed over enough items to be worth 1/3 of what remains (as opposed to 1/4 of the value of all the items).

Since these three players think more than 3/4 of the value is left after the first player departs with his or her goods, each of these three will think he or she gets more than (1/3)(3/4) = 1/4 of the total value. This continues, so that the players exiting later in the process have an obvious advantage over those exiting earlier. This bias favoring later players may tempt all players into deviating from the strategy of being honest about when they think the knife has reached the $1/n$ point.

A different version of this procedure was proposed in a National Science Foundation Chatauqua-type Short Course for College Teachers in 1975 by William F. Lucas (Lucas, 1994), which we refer to as *Lucas' method of markers*, or simply *Lucas' method.*[7] It has the advantage of forcing players to make their declarations about where the $1/n$ points in the list are before they gain any information about where others think these points are. This prevents certain lucky players (i.e., those who call stop later in the moving-knife procedure) from having a strategy that guarantees them strictly more than their proportional shares.

For Lucas' procedure to apply, an important assumption, called "linearity," must hold:

Linearity assumption: Players can create an equal division by placing markers at points on a straight line, along which the items to be divided are arrayed from left to right.

These markers divide the line up into n consecutive segments (i.e., equal to the number of players), with each marker – placed between two items – indicating the end of one segment and the beginning of another.

The linearity assumption precludes building equal piles by putting together *any* two goods. If there are two players, for example, each player must, in order to create equal nonempty consecutive segments, place a marker somewhere along the line. Now wherever the marker is placed, one pile cannot contain both the good at the left end and that at the right end of the line, because that would imply that one pile contains all the goods and the other none, which we assume is impossible.

Practically speaking, it seems highly unlikely that a division of discrete goods into exactly equal piles would ever be possible. This is because the most even possible division would probably leave some piles worth more than others, given that the players cannot divide any individual item in order to smooth out differences in the piles.

For applications, however, one can interpret the linearity assumption to mean that players are able to create a rough kind of equality. Moreover, in practice this roughness will often not matter – the piles will be close enough in value to be *considered* equal.

[7] Some ideas underlying Lucas' method, but not a formal procedure, can be found in Steinhaus (1969, pp. 70–1), with an application to land division. The name, "method of markers," was given to the procedure in Tannenbaum and Arnold (1992, pp. 81–5), but the procedure might better be characterized as the "method of equal division" for reasons that will soon become apparent.

To be sure, if there are divisible as well as indivisible goods to be divided, the problem of equalization is diminished. But we shall concentrate on the harder practical problem of allocating indivisible goods, though the procedure can also be applied to divisible goods, or a mixture of divisible and indivisible goods.

To illustrate the method of markers, assume there are three players, Bob, Carol, and Ted, who must divide up a set of nine goods, $\{a, b, c, d, e, f, g, h, i\}$. Assume the goods are arranged in alphabetical order.

To motivate the divisions by the players, assume Bob cares only about vowels and so places his markers, which we indicate by vertical lines, after each of the vowels in the list (there are three in all):[8]

Bob: $a \mid b\ c\ d\ e \mid f\ g\ h\ i$.

Carol cares only about consonants and divides up the line so that each segment contains exactly two consonants (there are six in all):

Carol: $a\ b\ c \mid d\ e\ f \mid g\ h\ i$.

Finally, Ted wants both a vowel and at least one consonant and so places his markers at points which give such a division:

Ted: $a\ b \mid c\ d\ e\ f \mid g\ h\ i$.

Except for Carol, the placement of markers is not uniquely determined by the above-postulated goals; we shall explore an alteration in these goals later.

We assume that the players independently place their markers along the line. (They might do so by indicating their placements to a referee, who then physically puts down all the markers at once.) We next illustrate, using the foregoing example, how Lucas' method ensures each player of a proportional allocation, given the linearity assumption.

It is useful to show all the markers for the three players in one picture, where B = Bob, C = Carol, and T = Ted:

$a \mid b \mid c \mid d\ e \mid f \mid g\ h\ i$.
B T C B C/T

If there are n players, each of whom is placing $n - 1$ markers along the line, then there are at most $n(n - 1) = n^2 - n$ markers all together, and so the line is divided into a maximum of $n^2 - n + 1$ different segments. In our example, however, there are not $3^2 - 3 + 1 = 7$ different segments but six, because the last (visible) markers of both Carol and Ted's partitions are the same. The

[8] We may think of there being a hidden marker at the end of the line – after the last vowel, i – so no marker need be placed there. In general, to partition a line into n consecutive segments, each player must place $n - 1$ markers along it.

partitions of the *entire* line by each player are, nevertheless, different in our example.

To apply the method of markers, we start from the left (we could as well start from the right, as we will illustrate later). We assign the first segment to the player who placed the first marker we reach as we scan from left to right. Thus, Bob gets $\{a\}$, which he regards as a proportional allocation (i.e., equal to 1/3, just as his second segment from b to e, and his third segment from f to i, are also equal to 1/3). Bob and $\{a\}$ are now removed from the line, along with his markers.

Notice that the segment removed does not touch any player's second or third segment. Thus, each of the remaining players (Carol and Ted) still has his or her second and third segments intact. At this point we have the following division:

$b \mid c \mid d\,e\,f \mid g\,h\,i.$
T C C/T

Again we scan from left to right until we come to a *second* marker of the remaining players, which in our example belongs to both Carol and Ted and occurs after the letter f. Moving backward from right to left to find which of these players' *first* markers we encounter first, it is Carol's, which follows c.

We give the letters between Carol's first and second markers $\{d, e, f\}$, to her, which she regards as a proportional allocation. Notice that the segment removed does not overlap Ted's third segment. Carol now leaves the game, and her markers, along with d, e, and f, are removed from the line, and the remaining player (Ted) has his third segment intact. At this point, we are left with the following division:

$b \mid c \mid g\,h\,i.$
T T

We now look for the *third* (hidden) marker of the only remaining player, Ted, which is the endpoint after i. Going back to find Ted's *second* marker, which follows c, we see that his last two markers encompass $\{g, h, i\}$. We give these letters to him, which he regards as a proportional allocation.

Altogether, Bob, Carol, and Ted receive the three sets of letters shown below in braces:

$\{a\}\ b\ c\ \{d, e, f\}\ \{g, h, i\}.$
Bob Carol Ted

The two letters, b and c, are, at this point unassigned. Since Bob has a vowel, Carol has two consonants, and Ted has a vowel and a consonant, however, they have all achieved their goals – and a proportional allocation – without these letters.

The method of markers works equally well if we start the process from the right rather than the left. In our example, this reversal gives two different allocations, depending on whether we assign the first segment on the right, $\{g, h, i\}$, to Carol or Ted. If Carol gets the first assignment, then Ted would get $\{c, d, e, f\}$ and Bob would get $\{a\}$, with only b left over. If Ted gets the first assignment, then the allocation would duplicate the allocation we obtained by going from left to right. In either case, or choosing any other order, the players receive proportional shares.

The goals we attributed to the players, based on their preferences for vowels or consonants, were simply meant to explain why they placed the markers where they did. To be sure, one can go backwards and infer something about players' goals from where they place their markers. But this will generally be an imprecise exercise and not give a definite indication of which players, if any, should receive what unallocated goods.

One could, of course, apply the method of markers again to the unallocated goods. But then, as in our example, there may be more players (3) than there are discrete goods left over to allocate (2 or 1).

This circumstance will generally prevent the players from making a second equal three-way division in placing their markers. Only if enough players have no desire for the leftover goods would an equal allocation be possible. Thus in our example, because Bob, by assumption, has no desire for consonants, and there are two left over if we go from left to right, we can give each of the other two players one each without upsetting Bob.

To summarize, given the linearity assumption is met, the method of markers is an n-person proportional procedure applicable to the allocation of indivisible goods, although the inclusion of divisible goods may facilitate the equal division. The placement of markers becomes the basis for making an allocation of the indivisible goods.

The method works. In particular, notice that the first segment allocated is a subset of everyone's first segment. Hence, at the second stage, we can be sure that the segment allocated is someone else's entire second segment, and that this allocated segment is contained within the first two segments of everyone left in the game. Then, at the third stage, we can be sure that the segment allocated is someone else's entire third segment, and that this allocated segment is contained within the first three segments of everyone left in the game. And so on.

In this manner, each player is assured that he or she can eventually receive one of his segments set off by markers. In fact, as our example illustrates, there usually will be leftover goods, allowing proportionality to be satisfied even without making a complete assignment of the goods.

Because the linearity assumption is unlikely to be met in practice, we purchase proportionality at a high price. In our example, if one of the goods (letters) is worth more than 1/3 for any player, then there is no way of dividing the remaining eight goods into two piles so that each is worth 1/3 each. Even if the most valuable good for a player is worth, say, only 1/5, if it comes first in the linear order followed by a second good worth 1/6, these two, which must be grouped together, total approximately 0.37, rendering impossible an equal division of the kind required.

The method of markers seems most appropriate if there are many small items and no particularly large ones, and the two are quite interspersed along the line. Then the kind of rough equality we spoke of earlier will be obtainable.

Some orderings of the items will work better than others in facilitating equal groupings by the players, but it may not be apparent *a priori* what these are. Therefore, it would be useful to have an algorithm, which would depend on how the players value the different goods, that could find a "best" ordering, perhaps minimizing the difficulty of the players in placing their markers or perhaps leaving a maximum surplus.

The order in which the goods are arrayed from left to right will, of course, affect which proportional segment a player receives. So will the direction in which the procedure is applied – from left to right or right to left. Because all players receive proportional shares under the method, and these are constructed from smallest segments, it might seem that this method is envy-free. This, unfortunately, need not be the case, regardless of how the leftover goods are assigned, as we will illustrate next.

Consider our earlier example, in which Carol received $\{d, e, f\}$, going from left to right. Although Bob received $\{a\}$, which is worth 1/3 to him, he may well consider $\{d, e, f\}$ worth more than 1/3 – even as much as 2/3 if he places little or no value on b and c in his second segment, and g, h, and i in his third segment. Hence, Bob may envy Carol, so the procedure is not envy-free.

3.4 Efficiency and entitlements

Recall that in section 2.8 we saw that all of the n-person proportional procedures in the divisible case turned out to be (in fact, were doomed to be) inefficient. Here the situation is different. Knaster's procedure is, in fact, efficient. The easiest way to see this is to observe that among all allocations of the items and money (subject to the requirement that money simply gets shifted from some players to other players), Knaster's procedure maximizes the sum – over all the players – of the monetary value that each player perceives he or she received in items and money.

The reason Knaster's procedure maximizes this "total welfare" is that each player is receiving the items he or she values more than any of the other players. (The compensation paid by some players to others does not affect the total welfare.[9]) Now it is easy to see that if some other allocation were better for everyone than the allocation produced by Knaster's procedure, then that other allocation would yield a larger total welfare, contradicting the maximality of total welfare that Knaster's procedure produces.

Lucas' method of markers, however, does not automatically yield an efficient outcome, because it requires only an incomplete specification of the preferences of players (only indifference sets). Although each player in our example in section 3.3 ended up with exactly 1/3 of the items – as he or she valued them – there is no guarantee that this outcome is efficient. In particular, assume that Bob prefers $\{d, e, f\}$, which overlaps two of his indifference sets, to $\{a\}$, so give him $\{d, e, f\}$. Give Ted $\{g, h, i\}$, which he values at 1/3, and $\{a, b, c\}$ to Carol, which she values at 1/3.

This new allocation is, according to our assumption, better for Bob and the same for Carol and Ted as that given by the method of markers, rendering the allocation given by this method inefficient. Nonetheless, the method may be useful in dividing up a large number of small items, in which everybody can create more or less equal groupings, even if it does not always produce an efficient outcome.

Both Knaster's procedure and the method of markers can handle entitlements via the same technique introduced in section 2.8 for the other proportional procedures. For example, assume that the will in our example for Knaster's procedure in section 3.2 stipulates a 1/2 share to Bob, a 3/8 share to Carol, and a 1/8 share to Ted. We could, as before, pretend there are four Bobs, three Carols, and one Ted, and use Knaster's procedure as it applies to eight players. A more straightforward approach to this, however, is to calculate the initial fair shares as 1/2 x \$13,300 = \$6,650, 3/8 x \$8,500 = \$3,187.5, and 1/8 x \$14,000 = \$1,750 instead of the figures shown in table 3.1.

Although the players would receive the same items, the excess of Bob, the deficit of Carol, and the excess of Ted would be different from those given in table 3.1. (For example, Ted's initial excess would be greater because, after receiving items B and C, he would have to make a larger contribution to the surplus because of his now smaller entitlement.) These changes would lead to

[9] Other reasonable ways to share the surplus generated by Knaster's procedure and variants include the maximin procedure of Dubins (1977) and procedures discussed in Moulin (1988a, chapters 6 and 8) and Young (1994, pp. 131–4). Kleinman (1994) proposes a new procedure that requires that the players make not only sealed bids, which fixes the price of each item (defined as the average of all the bids), but also cash bids, which together with the sealed bids determine who wins each item (not necessarily the player who makes the highest sealed bid if his or her cash bids are too low). While Kleinman's procedure tends to bring players' sealed bids into line with their willingness to pay for items – as determined by their cash bids – it is not efficient if a player's cash bids, in the end, are not sufficient to pay for the items he or she wins. Young (1994, pp. 131–4) discusses this problem and proposes still a different auction procedure, which also is not efficient.

a different surplus, which the players would divide according to their different entitlements to determine new adjusted fair shares and new final settlements.

In chapters 6 and 7 we will analyze and compare envy-free procedures for three or more players. As we will show, they in general do not yield efficient outcomes. But first we revisit two-person fair-division procedures in chapters 4 and 5 that are not only proportional and, consequently, envy-free but also equitable (in a sense to be described); in addition, one of these procedures is efficient. This procedure, which is a point-allocation procedure quite different in spirit from the procedures discussed so far, is perhaps the most practical of the two-person procedures we analyze. Indeed, we recommend its use in the settlement of everything from divorces to international conflicts.

4 Envy-freeness and equitability for $n = 2$

4.1 Introduction

As we saw in chapters 2 and 3, proportional fair-division procedures do not generally produce efficient fair divisions, even in the two-person case (Knaster's procedure, described in section 3.2, is an exception). The same will turn out to be true for several envy-free procedures we will describe in chapters 6 and 7. In this chapter, however, we will show that envy-freeness and efficiency need not be incompatible, at least when there are only two players.[1]

The framework we use, however, is very different from that of the previous chapters. Instead of starting with a heterogeneous cake or set of indivisible items, we begin with *k* discrete goods (or issues), each of which we assume is divisible.[2] It is these that must be divided between two players.

In this new context, the two players explicitly indicate how much they value each of the different goods by distributing 100 points across the goods.[3] This information, which may or may not be made public, becomes the basis for making a fair division of the goods.

Valuations in cake division, by contrast, are implicit in the procedures used. They determine how the players divide a cake into equal pieces, which ones they judge to be acceptable, and so on. But the players are never required to say explicitly how much they value the different parts of the cake.

[1] This chapter is adapted from Brams and Taylor (1994b).

[2] We shall use goods and issues interchangeably in the subsequent analysis. "Goods" are more appropriate when discussing fair-division problems, like a divorce settlement between a husband and wife on how to split or share the property. "Issues" are more appropriate when analyzing negotiations, such as between labor and management on how much each side is willing to concede on the different issues that separate them.

[3] In the mid 1980s, Leng and Epstein (1985) and Salter (1986) proposed point-allocation schemes to facilitate superpower arms reductions, which were bogged down at the time. Under the Leng–Epstein proposal, each superpower would distribute, say, 100 points over its adversary's weapons; the adversary would then have to destroy weapons that would reduce these points by a particular percentage. Under the Salter proposal, each superpower would distribute points over its own weapons; the adversary would then effect a particular percentage reduction in these self-assigned points.

Requiring that players assign points to different goods raises the question of whether they will have an incentive to be honest in announcing their valuations. Likewise, announcing how important one considers issues in a dispute raises the question of whether candor is consistent with good bargaining tactics.

The role of candor or honesty will be central in our analysis of two procedures, and a combination of the two, in this chapter. Young (1994, p. 130) pinpoints why honesty is a problem:

> Preferences are usually private information, and we cannot expect people to honestly reveal them unless it is in their interest to do so. The challenge, therefore, is to design procedures that *induce* the claimants to reveal enough information about their preferences so that an equitable and efficient solution can be implemented [italics in original].

We begin by indicating how other theorists have incorporated bargaining tactics, and the underlying incentive of players to misrepresent their preferences, into the study of dispute resolution. We also discuss the means they propose to ameliorate this problem. We then describe each of our two procedures, analyzing the properties that each satisfies.

We also consider how they might be combined, using one as a "default option" if either player considers the allocation provided by the other procedure unacceptable. Finally, we show that none of our procedures can be extended to conflicts with more than two players without giving up some desirable feature, which we discuss in the context of land division, both past and present.

4.2 Bargaining and fair division

By and large, bargaining theory has proved singularly inapplicable to the settlement of real-life disputes. This is true despite the attempts made by a number of theorists to demonstrate the contributions that rational-choice models have made to understanding real-life conflicts and prescribing solutions (Raiffa, 1982; Lax and Sebenius, 1986; Brams, 1990b).

One reason for this failure, in our opinion, has been the divorce of bargaining theories – and, on the more applied side, "negotiation analysis" (Young, 1991; Sebenius, 1992) – from theories of fair division.[4] In this chapter and the next, we offer a reconciliation of these different theoretical strands by introducing two different fair-division procedures (chapter 4). In

[4] Recent exceptions to this generalization include van Damme (1991, chapter 7), Rabin (1993), Roemer (1994, part III), and Young (1994); we will distinguish our work from some of this work in section 4.3.

chapter 5 we apply them to the resolution of multiple issues in the case of the Panama Canal treaty dispute in the 1970s, and the division of multiple items in a hypothetical and in a real divorce case.

We evaluate the two procedures introduced in this chapter with respect to two already familiar criteria: (1) envy-freeness and (2) efficiency. We also introduce two new normative criteria: (3) the degree to which the procedures induce bargainers to be truthful, or at least almost truthful, which we briefly alluded to in connection with Knaster's procedure (section 3.2), and (4) equitability (which we shall define shortly).

Both fair-division procedures satisfy the properties of (1) envy-freeness (no surprise, because there are only two players) and (4) equitability (a surprise), and each also satisfies one of the other criteria. Neither procedure, however, satisfies all four criteria: there is a trade-off between (2) efficiency and (3) truthfulness, with one procedure satisfying (2) and the other satisfying (3).

The first procedure is *Adjusted Winner* (AW); it produces a settlement that is efficient, envy-free, and equitable with respect to the bargainers' *announced* preferences. However, because AW provides little inducement for the bargainers to be truthful in announcing their preferences, it may produce settlements that only *appear* to satisfy these criteria because (3), truthfulness, is not also satisfied.

For this reason, we propose a second fair-division procedure, called *Proportional Allocation* (PA), which could provide a "default" settlement should either party object to the settlement under AW. That is, PA could be implemented if either party, feeling that it was exploited under AW because of AW's vulnerability to false announcements, requests PA. For reasons to be discussed later, however, we think this safeguard will hardly ever be necessary.

Like AW, PA is envy-free and equitable; unlike AW, it is extremely robust against false announcements in most situations, thereby inducing the bargainers to be truthful. However, the settlement it yields is not efficient, so (2) is not satisfied. Nevertheless, it is substantially better for both parties than the naive fair-division procedure of splitting every issue 50–50.

Keeney and Raiffa (1991), in the absence of a procedure for ensuring an efficient settlement, propose that the parties to a dispute first work out an "acceptable" settlement, though they leave vague what this means. They suggest that a third party ("contract embellisher") might then make adjustments in the original settlement that moves it toward efficiency (again without saying exactly how) in what Raiffa (1985, 1993) calls a "post-settlement settlement."

By contrast, AW guarantees efficiency, as well as envy-freeness and equitability, at the start; the issue for the parties is whether it is "safe" to buy into a procedure that can, in principle, be exploited. This is precisely why we raise the question of using PA as a default option.

Although we think AW obviates the need for the haggling phase of negotiations – even if facilitated by a neutral third party – that Keeney and Raiffa (1991) recommend, our framework is similar to theirs. There are two parties and k issues ($k \geq 2$) that need to be resolved. Each party can quantify the relative importance of each issue to itself by distributing a total of 100 points over the k issues. Moreover, for each issue there is a set of unambiguously stipulated "resolution levels" which, as Keeney and Raiffa (1991) point out, may be either finite or a continuum.[5]

In an economic context, the problem we consider here is equivalent to that of dividing k infinitely divisible homogeneous goods between two consumers who value the goods differently. A *homogeneous good* corresponds to our earlier "issue for which a set of unambiguous resolution levels has been stipulated," and *infinite divisibility* corresponds to the resolution levels' being a continuum (assumed previously for cake cutting). Thus, "Bob gets 60 percent of good i" corresponds to "issue i's being resolved 60 percent in favor of Bob and 40 percent in favor of Carol."

More formally, given k goods, $G_1, \ldots, G_k$, and two players, Bob and Carol, we assume both players can independently assign points to the goods that indicate their true value for each. We also assume that the goods are "separable" and the points "additive," whose meaning we will discuss after describing the first procedure (AW) in section 4.3. Assume that Bob's true values are $a_1, \ldots, a_k$ and Carol's $b_1, \ldots, b_k$, where the a_i's and b_i's are positive and sum to 100.

The a_i's and b_i's may or may not be *common knowledge*, whereby each player knows these values, knows that the other player knows them, and so on *ad infinitum*. In either case, we assume that $x_1, \ldots, x_k$ and $y_1, \ldots, y_k$ are the players' *announced points*, which may or may not reflect their true valuations. Our interest is in dividing each of the goods between the two players so that the resulting allocation is satisfactory, according to some – if not all – the aforementioned criteria.

4.3 The Adjusted-Winner (AW) procedure

AW allocates k goods as follows. Let X be the sum of the points of all goods that Bob announces that he values more than Carol does. Let Y be the sum of the values of the goods that Carol announces she values more than Bob does. Assume $X \geq Y$. Next, assign the goods so that Bob initially gets all the goods where $x_i \geq y_i$, and Carol gets the others. Now list the goods in an order G_1, G_2,

[5] In fact, as we will show later, AW requires that only one issue be divisible, though which one this is will not be known at the start. In the subsequent analysis, we assume that all issues are divisible, but we will reconsider this assumption in section 4.5, where we discuss how AW might be used in negotiations in which some issues are indivisible.

etc., so that the following hold:

1 Bob, based on his announcement, values goods $G_1, \ldots, G_r$ at least as much as Carol does (i.e., $x_i \geq y_i$ for $1 \leq i \leq r$), where $r \leq k$.

2 Carol, based on her announcement, values goods $G_{r+1}, \ldots, G_k$ more than Bob does (i.e., $y_i > x_i$ for $r + 1 \leq i \leq k$).

3 $x_1/y_1 \leq x_2/y_2 \leq \ldots \leq x_r/y_r$.

Thus, Bob is initially given all goods 1 through r that he values at least as much as Carol, and Carol is given all goods $r + 1$ through k that she values strictly more than Bob.

Because $x_i \geq y_i$ for $1 \leq i \leq r$, the ratios in (3) are all at least 1. Hence, all the goods for which $x_i = y_i$ come at the beginning of the list. Bob – who, because $X \geq Y$, enjoys an advantage (if either player does) after the winner-take-all assignment of goods – is helped additionally by being assigned all goods that the players value equally, based on their announcements.

The next step involves transferring from Bob to Carol as much of G_1 as is needed to achieve *equitability* – that is, until the point totals of the two players are equal.[6] (Recall that equitability is only apparent, not true, because we do not assume that the players' announcements of their point assignments are necessarily truthful.) If equitability is not achieved, even with all of G_1 transferred from Bob to Carol, we next transfer G_2, G_3, etc. (in that order) from Bob to Carol. As we will illustrate shortly, it is the order given by (3), starting with the smallest ratio, that ensures efficiency.[7]

Example

Suppose there are three goods for which Bob and Carol announce the following point assignments (the larger of the two assignments is underscored):

	G_1	G_2	G_3	Total
Bob's announced values	$\underline{6}$	$\underline{67}$	27	100
Carol's announced values	5	34	$\underline{61}$	100

6 Austin's moving-knife procedure (section 1.5) also achieves equitability by giving two players pieces of size exactly 1/2 (or $1/k$). Unlike AW, however, Austin's procedure does not enable the players to receive more, by the same amount, than 50 percent each, as AW does. Our usage of the term "equitability" is different from Varian's (1974) use of "equity," which means what we call envy-freeness. In Feldman and Kirman (1974) and Crawford (1977), envy-free allocations are called "fair," and in Baumol (1986) "superfair." Some of these terminological differences are discussed in Thomson and Varian (1985), which provides an overview of the earlier economic literature. Tie-ins of economics and game theory to the philosophical literature on justice and egalitarianism are given in Binmore (1994), Roemer (1994), and Kolm (1995).

7 What we are calling "equitability" dates back to Pazner and Schmeidler's (1978) concept of "egalitarian equivalence." Their concept, which is ordinal and does not require interpersonal comparisons (more on this later), finds an allocation for which there is a number λ so that each player is indifferent between what he or she receives and receiving the fraction λ of each good. This makes "the common property itself (rather than money) as the standard" (Young, 1994, p. 148); in our context, point totals are equalized when there exists such a λ.

Initially, G_1 and G_2 are assigned to Bob, giving him 73 of his points, and G_3 is assigned to Carol, giving her 61 of her points. Hence, goods must be transferred from Bob to Carol to create equitability.

Notice that $x_1/y_1 = 6/5 = 1.2$ and $x_2/y_2 = 67/34 \cong 1.97$, so the smallest ratio of the players' valuations is for G_1. Even transferring all of G_1 from Bob to Carol, however, still leaves Bob with an advantage (67 of his points to 66 of hers).

Let α denote the fraction of G_2 that will be retained by Bob, with the rest transferred from him to Carol. We choose α so that the resulting point totals are equal for Bob (left side of the equation) and Carol (right side of the equation):

$$67\alpha = 5 + 34(1 - \alpha) + 61,$$

which yields $\alpha = 100/101 \cong 0.99$. Consequently, Bob ends up with 99 percent of G_2 for a total of 66.3 of his points, whereas Carol ends up with all of G_1 and G_3, and 1 percent of G_2, for the same total of 66.3 of her points.[8] We call this the *equitability adjustment*, which equalizes the number of points both players possess.

AW has three compelling properties, two of which are obvious by construction and one of which is not. The properties are given in theorem 4.1, which we prove in the appendix to this chapter but for which we provide an informal rationale here. We then offer some comments on the meaning of the properties themselves.

Theorem 4.1 AW produces an allocation of the goods, based on the announced values, that is:

(1) efficient: *any allocation that is strictly better for one player is strictly worse for the other;*

(2) equitable: *Bob's announced valuation of his allocation is the same as Carol's announced valuation of her allocation;*

[8] The reader can check that if the first transfer had been part of G_2 instead of G_1 to Carol, both players would have received 65.0 of their points and hence have fared worse. How transferring part of an indivisible good, like a house, might be accomplished will be discussed later. For now we note that the transfers never require the splitting or sharing of more than one good (G_2 in this case, the last good transferred). We also note that AW can be modified to reflect unequal shares to which the parties might be entitled (e.g., 2/3 to Bob, 1/3 to Carol), which in our example can be accomplished by giving Bob both G_1 and G_2 and an α proportion of G_3, with a $1 - \alpha$ proportion of G_3 going to Carol. To ensure that Bob receives twice as many points as Carol does, we set his points (left side of equation) equal to twice her points (right side of equation in brackets):

$$67 + 6 + 27\alpha = 2[61(1 - \alpha)].$$

Solving for α yields $\alpha \cong 0.33$, which gives Bob 81.8 points (15.1 points above 66.7) and Carol 40.9 points (7.6 points above 33.3). Hence, both players receive about 23 percent more points than their entitlements, ensuring equitability, which is somewhat less than the 33 percent gain (16.3 points above 50) that each realizes when their entitlements are 50–50.

(3) envy-free: *neither player would trade his or her allocation for that of the other.*

Efficiency depends on showing that there can be no better allocation for both players. This is nontrivial and requires a somewhat extended mathematical argument. There are two things that make the argument work: (1) we start with an efficient distribution (giving each player all the goods he or she most values); and (2) we make the equitability adjustment in the prescribed order (i.e., based on the smallest-ratio criterion). Using this line of attack, one can then show that it is impossible to help both players with a different allocation, including one that is not equitable.

Equitability – the new property in this chapter that we shall discuss shortly – is built into AW by construction (i.e., through the equitability adjustment of the points of the two players). Envy-freeness follows from the fact that each player will wind up with at least 50 of his or her points, although this is not completely obvious and requires a technical argument.

Note that envy-freeness and equitability both address the question of whether one player believes he or she did at least as well as the other player. The difference is that envy-freeness involves an internal comparison, based on a player's own valuation, which is captured by the following question:

Are you at least as well off with your allocation as you would be with your opponent's allocation and, hence, would not desire to trade with your opponent?

Equitability, on the other hand, involves a more controversial external or interpersonal comparison (Elster and Roemer, 1991), which is captured by the following question:

Is your announced valuation of what you received equal to your opponent's announced valuation of what he or she received?

In other words, did you receive, according to your point assignment, exactly what your opponent received according to his or her point assignment?

Equitability, however, need not *directly* involve the comparison of one player's valuation of his share with the other player's valuation of her share. If there is a fraction λ of all the goods being allocated (e.g., 2/3 of each good), and each player is indifferent between receiving this fractional allocation and the allocation that he or she actually received, then the players' allocations are equitable. Pazner and Schmeidler (1978) called such allocations "egalitarian equivalent" (see note 7); because the fraction λ is common to all the players, the allocations are egalitarian (equal).

Because we allow for different entitlements, we prefer the term "equitable" – echoing Pazner and Schmeidler's (1978) subtitle of their article, "a new concept of equity" – to describe situations in which the players do not necessarily receive equal allocations, in terms of their own valuations. What they do receive, whether their entitlements are the same or different, are allocations that are the same percentage greater than their entitlements (see note 7 for an example with different entitlements).

Again, what we are calling "envy-freeness" and "equitability" are only "apparent envy-freeness" and "apparent equitability" if the players are not truthful. When the players are truthful – $x_i = a_i$ and $y_i = b_i$ for all i – in the equal-entitlements case, which we shall henceforth assume, each player *assuredly* receives at least 50 points (based on his or her own valuation), and the surplus above 50 points is the same for each (i.e., his or her "more" is the same as his or her opponent's "more").

The equitability adjustment that gives each player 66.3 of his or her points in our example may be interpreted as providing each player with nearly 2/3 of what he or she perceives to be the total value, or utility, of all three goods.[9] This equalization of the players' utilities assumes that points (or utilities) are additive and linear. *Linearity* here means that the players' marginal utilities are constant – instead of diminishing as one obtains more of something – so, for example, $2x$ percent of G_i is twice as good as x percent. (This usage of linearity differs from its usage for Lucas' method of markers in section 3.3.) *Additivity* here means that the value of two or more goods to a player is equal to the sum of their points (this usage is the same as our usage for Knaster's procedure of sealed bids in section 3.2).

Neither assumption is necessarily a good reflection of players' preference functions on certain issues, which is a matter we return to in section 5.2. Thus, goods may not be "separable" because of complementarities – that is, obtaining one good may affect the value one obtains from others.

Perhaps the main drawback of AW is the extent to which it fails to induce the players to be truthful about their valuations – and thereby fails to lead to an envy-free, equitable, and efficient outcome, based on these valuations.[10]

[9] We shall not discuss here how, in the abstract, one might determine one's utility for each good, which essentially involves determining when one would be indifferent between receiving the good and having some specified chance, based on a lottery, of receiving certain other goods. Utilities determined in this way are called von Neumann–Morgenstern utilities and meet certain conditions; see, for example, French (1986) and Keeney (1992). More concretely, we shall in chapter 5 give examples of utility assignments that players have made or might make to issues or goods in a dispute.

[10] This is a general problem with mechanism design (Osborne and Rubinstein, 1994, chapter 10): players' truthful revelations of their valuations do not constitute a Nash equilibrium (to be defined shortly in the text and then illustrated for AW and PA). However, there are (non-standard) bargaining procedures that induce, in equilibrium, truthful revelation via dominant strategies (Brams, 1990b, chapter 2; Brams and Kilgour, 1995). In fact, honesty can always be induced, according to the "revelation principle," by so-called direct mechanisms (Myerson, 1979; 1991, chapter 6; Fudenberg and Tirole, 1991, pp. 253–7; Binmore, 1992, pp. 530–2; Rasmusen, 1994, pp.198–9), but these mechanisms do not in general lead to efficient outcomes (Tadenuma and Thomson, 1995). (As we will show later, AW is efficient but not honesty-inducing, whereas PA is inefficient but relatively honesty-inducing.)

This is easy to illustrate, even in the case of two goods. Suppose Bob values the goods equally, and Carol knows that he will truthfully announce his 50–50 valuation. Suppose Carol's true valuation is 70–30. What should she announce? Assuming that announcements must be integers, the answer is 51–49, as we will show shortly.

The result of this announcement will be an initial allocation of all of G_1 to Carol (which she values at 70), and all of G_2 to Bob (which he values at 50). Then there will be a transfer of only a trivial fraction (1/101) of G_1 to Bob, since it appears that Carol's initial advantage is only 51 of her points to 50 of Bob's points. Thereby Carol will end up with a generous $70 - 0.7 = 69.3$ points (according to her true valuations), but Bob will realize only $50 + 0.5 = 50.5$ points (according to his true valuations).[11]

Bob can turn the tables on Carol if he knows her values of 70–30 and that she will announce these. If Bob announces 69–31, there will be a transfer of 39/139 of G_1 from Carol to Bob, giving him a total of $50 + 14.0 = 64.0$ points and her only $70 - 19.6 = 50.4$ points, based on their true valuations.

Thereby one player (with complete information) can exploit another player (without such information). On the other hand, if both players were truthful in their announcements, there would be a transfer of 1/6 of G_1 from Carol (70–30) to Bob (50–50), giving each player 58.3 points.

If the announced and real values are restricted to integers, then optimal responses and *Nash equilibria* (Nash, 1951) – outcomes from which neither player would have an incentive to depart unilaterally because his or her departure would lead to a worse, or at least not a better, outcome – can easily be computed.[12] In fact, the determination of whether a response is optimal requires only 100 comparisons, based on the answer to the following question:

[11] It is easy to see that if announced valuations are not restricted to the integers, then there is *no* optimal response for Carol in this example, because the payoffs are discontinuous. That is, if $P(y)$ is the payoff to Carol resulting from her announced valuation of $50 + y$ for G_1, then $P(y)$ approaches 70 as y approaches 0, even though $P(0)$ has an expected value of 50 (because it is 30 half the time and 70 half the time).

[12] AW can always be implemented as a Nash equilibrium using Crawford's (1979) procedure of bidding to be the divider; see also Demange (1984) and Young (1994, pp. 143–5, 159). But in order for the players to know what to bid, implementation requires that their preferences be common knowledge (section 4.2), which seems highly unlikely. AW, on the other hand, operates in a context in which each player knows only his or her own preferences. As we will show shortly, however, once the players' assignments of points are revealed, either may have an incentive to deviate – were they to be given a chance – so their strategies of assigning points according to their (true) preferences may not be a Nash equilibrium. In chapters 8 and 9, equilibrium solutions will be proposed to different games that do not require the play of the rather demanding bidding games of Crawford (1979) and Demange (1984), which are criticized in Sjöström (1994). Cooperative game-theoretic concepts, such as the core, have also been applied to fair division; see Legut (1986, 1987, 1988a, 1990) and Legut, Potters, and Tijs (1994). Overviews of some of the cooperative solutions can be found in Moulin (1988a) and Young (1994).

Holding the other player's allocation to G_1 and G_2 fixed, would the player in question do better choosing any *other* of the 100 integer allocations to G_1 (and the complementary allocation to G_2) than the one in question? If not, then his or her present response is optimal.

Testing a pair to see if it is a Nash equilibrium requires at most 198 comparisons, based on the answers to the following question: Would either player do better – holding the other player's allocation fixed – by choosing any other of the 99 integer allocations than either (1) the one in question or (2) the other player's allocation? In the case of (2), its choice would result in a tie, and each player's receiving exactly 50 points, which is an outcome that can only be improved upon, so it cannot be an equilibrium.

But such computations are not really needed, because the following theorem, proved in the appendix to this chapter, and its corollary completely settle the question of optimal responses and Nash equilibria in the integer-valued case:

Theorem 4.2 Assume there are two goods, G_1 and G_2, all true and announced valuations are restricted to the integers, and suppose Bob's announced valuation of G_1 is x, where $x \geq 50$. Assume Carol's true valuation of G_1 is b. Then her optimal announced valuation of G_1 is:

$$\begin{array}{ll} x+1 & \text{if } b > x \\ x & \text{if } b = x \\ x-1 & \text{if } b < x. \end{array}$$

Example

In our earlier example, $x = 50$ and $b = 70$, so Carol's optimal announcement is $x + 1 = 51$. When the tables are turned and $x = 70$ and $b = 50$, Bob's optimal announcement is $x - 1 = 69$.

Corollary 4.1 Assume all true and announced valuations are restricted to the integers, and suppose Bob's true valuation of G_1 is b, and $a > b$. Then the Nash equilibria are the following ordered pairs of announced valuations for G_1 by Bob and Carol:

$$\begin{array}{lll} (x+1, x) & \text{where} & b < x < a-1; \\ (a, a) & \text{where} & a = b. \end{array}$$

Example

To ensure $a > b$ in our earlier example, let the 70–30 player be Bob, so $a = 70$, and the 50–50 player be Carol, so $b = 50$. Then the Nash equilibria are all ordered pairs $(x + 1, x)$ for Bob and Carol, respectively, where $50 < x$

< 69. That is, the 20 pairs

$$(51, 50), (52, 51), \ldots, (70, 69)$$

are precisely the announcements that Bob and Carol can make such that neither player would have an incentive to depart unilaterally from his or her announcement because of a worse outcome in the case of such a departure. Note that these announcements differ by only one point but do not include an honest announcement by both players – that is, (70, 50) by (Bob, Carol) for G_1.

4.4 The Proportional-Allocation (PA) procedure

PA, which we shall next describe, comes much closer to inducing the players to be truthful. Consider again our earlier example of exploitation with AW, wherein Bob (50–50) announced his true valuation, and Carol (70–30) – knowing Bob's allocation – optimally responded by announcing 51–49. Thereby, Carol obtained 69.3 points, compared with the 58.3 points that truthfulness would have given her (a 17.2 percent increase).

Under PA, as we will show, the optimal response of Carol is to be nearly truthful, announcing 71–29 instead of 70–30. Her benefit from this slight distortion of the truth is only in the third decimal place, gaining her 52.087 points compared to 52.083 points (less than a 0.01 percent increase). But note that both players do worse, when truthful, under PA (52.1 points) than under AW (58.3 points), so PA is not efficient.

Later we will weigh the relative nonmanipulability of PA against the efficiency of AW. A clear-cut choice is not necessary, however, if we use PA as a default option to AW, which – under the so-called combined procedure – either player can invoke if he or she feels exploited. Yet this complication, we will argue, is probably unwarranted in most practical applications of the procedures.

Although PA does not give an efficient allocation, like AW it is equitable – though this is by no means evident – and envy-free. As we just illustrated, it also comes remarkably close to inducing truthfulness, at least in situations where no good is of either negligible or of overriding value to either player.

PA, as its name implies, allocates goods proportionally. As before, assume that Bob announces values of $x_1, \ldots, x_k$, and Carol announces values of $y_1, \ldots, y_k$, for goods $G_1, \ldots, G_k$. Assume that for each i, either $x_i \neq 0$ or $y_i \neq 0$. Then Bob is allocated the fraction $x_i/(x_i + y_i)$ of G_i, and Carol the fraction $y_i/(x_i + y_i)$.

Example

Consider our earlier example of three goods, for which Bob and Carol announce the following point assignments:

Table 4.1. *Optimal responses of Bob to Carol's announced valuations under PA*

True valuation of Bob	Announced valuation of Carol						
	20	30	40	50	60	70	80
20	20	19.32	17.98	16.67	15.96	16.61	20
30	29.32	30	29.67	29.13	29.00	30	33.39
40	37.98	39.67	40	39.90	40	41.00	44.04
50	46.67	49.13	49.90	50	50.10	50.87	53.33
60	55.96	59.00	60	60.10	60	60.33	62.02
70	66.61	70	71.00	70.87	70.33	70	70.68
80	80	83.39	84.04	83.33	82.02	80.68	80

	G_1	G_2	G_3	Total
Bob's announced values	6	67	27	100
Carol's announced values	5	34	61	100

Bob is awarded 6/11 of G_1, 67/101 of G_2, and 27/88 of G_3, giving him a total of 55.9 of his points. Likewise, Carol also receives a total of 55.9 of her points. (Recall that AW awarded both players 66.3 points when they were truthful, or 18.6 percent more than PA gives in this example.) The equitability (as well as the envy-freeness) of PA is no accident:

Theorem 4.3 PA produces an allocation of the goods, based on the announced values, that is equitable and envy-free.

The proof of this theorem is given in the appendix to this chapter. We remark that it is surprising that PA, in helping each player gain proportionally more points on the goods it rates the highest – relative to the other player – enables each to garner exactly the same total number of points (when the points received on each good are summed across all the goods).

PA and AW do not exhaust the allocation procedures that are equitable and envy-free. For example, the naive procedure we alluded to in section 4.1 of splitting every good 50–50 gives each player exactly 50 points, so it satisfies both desiderata. Yet, not only is this allocation less efficient than AW (66.3 points for each player in our earlier example), but it is also less efficient than PA (55.9 points for each player).

The principal advantage of PA over AW is that it discourages departures from truthfulness of the kind we showed to be optimal under AW in section 4.3. When there are only two goods, this discouragement will be absolute under

Table 4.2. *Nash-equilibrium announcements under PA*

True valuation of Bob	True valuation of Carol			
	20	30	40	50
20	20/20	19/29	18/37	17/45
30	29/19	30/30	30/40	29/49
40	37/18	40/30	40/40	40/50
50	45/17	49/29	50/40	50/50
60	54/16	59/29	60/40	60/50
70	64/16	70/30	71/41	71/51
80	80/20	84/36	84/46	83/55

PA if and only if (1) the preferences of the players coincide or (2) they are diametrically opposed, as we will illustrate presently.

But what if the players are neither in complete agreement nor in complete disagreement? We next show that truthfulness is still a good – if not quite optimal – strategy under PA when there are two goods and the players do not attribute overriding value (i.e., more than 80 percent) to one good or the other.

To ascertain the incentive of a player to depart from truthfulness, we show in table 4.1 the optimal response of Bob (row), for his true valuations between 20 and 80, to Carol's (column's) announced valuations between 20 and 80. (We shall consider an example of more extreme valuations outside the 20–80 range shortly.) Note first that Bob should be truthful if his valuation is the same as, or diametrically opposed to, Carol's announced valuation.

When this is not the case, Bob's optimal response is generally very close to truthfulness. For example, if Carol's announced valuation is 50, and Bob's is 70, then Bob's optimal response turns out to be to allocate 70.87 points to this good. His payoff, as shown earlier, will then be 52.087, compared with 52.083 when he is truthful. In other words, his payoff is hardly affected (only in the third decimal place) by deviating in an optimal way from truthfulness.

In table 4.2 we give the Nash equilibrium announcements of Bob (row) and Carol (column) for true valuations between 20 and 80 points.[13] These valuations are rounded to the nearest integer and show that the players' equilibrium announcements in the 20–80 range do not differ very much from their true valuations.

[13] We do not show the valuations of Carol for 60, 70, and 80 points, because the results we do not present can be surmised from other entries in table 4.2. For example, if Bob and Carol's valuations of G_1 are 60 and 70, respectively, which are not shown in table 4.2, they must be 40 and 30 for G_2, which are shown.

For example, Bob and Carol's true valuations of 50 and 20 translate into 45/17 equilibrium announcements, which in relative terms is the most significant departure from truthfulness in the table. (Note that 50/20 is equidistant in the first column of table 4.2 from the two truth-inducing valuations of 20/20 and 20/80.) The payoff consequences, however, are quite small: 50/20 gives both players 54.95 points under PA, whereas 45/17 gives the 50-point player 56.22 points and the 20-point player 53.60 points.[14]

Consider a more extreme valuation than that given in table 4.2 – namely, 90–10 by Bob, and 10–90 by Carol, for G_1 and G_2, respectively. While truthfulness (90/10 and 10/90) is an equilibrium, it is nonequilibrium announcements of 100/0 by Bob and 0/100 by Carol that are efficient under PA, giving the players equitable and envy-free payoffs of 90 points each.[15] In fact, these are the payoffs that AW would also give: all G_1 would go to Bob, and all G_2 would go to Carol, with no equitability adjustment necessary.

We have shown that PA induces truthfulness, at least to a high degree, but it is purchased at the cost of the efficiency that AW gives. Is it possible somehow to combine these two procedures and get the best properties of both?

4.5 The combined procedure

In the absence of reliable intelligence about the announced valuation of the other player (perhaps obtained by spies), it seems likely that the players will stick with their true values under PA – especially in the 20–80 range – because there are equilibrium announcements "close" to these values. On the other hand, if a player's valuation exceeds 90 for one good, he or she may be tempted to put all 100 points on it – unless the other player has either exactly the same or exactly the opposite valuation and intends to announce it, in which case truthfulness on the part of both players is a Nash equilibrium.

But, as we just illustrated, Nash equilibria may not be efficient; by contrast, AW always guarantees efficiency when the players are truthful. For this

[14] The calculation of payoff consequences such as these shows that truthfulness is not literally a *dominant strategy* – that is, one that gives an outcome which is at least as good as, and sometimes better than, any other strategy, regardless of the strategy choice of one's opponent. However, for the range of values we are considering, payoffs for being truthful are within about 3 percent of being a best response to *any* choice of one's opponent. Consequently, truthfulness is a near-dominant strategy under PA even if, in the absence of complete information about an opponent's preferences, players cannot make the precise optimal-response or Nash-equilibrium calculations described in the text.

[15] When Carol announces 0/100, Bob can optimally respond with 1/99, which is why 100/0 and 0/100 does not constitute a Nash equilibrium when the true valuations of the players are 90–10 and 10–90. This optimal response on the part of Bob to Carol's announced valuation of 100/0 garners him nearly 95 points under PA: $(1/1)(90) + (99/199)(10) \cong 94.97$ points. By comparison, Carol receives $(0/1)(10) + (100/199)(90) \cong 45.23$ points, which illustrates how dishonesty by a player can bring his or her point total below 50 points.

reason, assume AW is used initially. Its vulnerability to manipulation, however, suggests that the players might be given the opportunity to opt for the strategically more robust PA, allowing players to revise their announcements, if at least one player so chooses. (There is also an argument for not allowing players to revise – we shall return to this later.)

This default option might well be selected in the kind of situation we illustrated in section 4.3, in which Bob announces 50–50 and Carol announces 51–49. If Bob thinks that Carol really values G_1 much more than 51 points (say, at 70 points instead), then he could invoke PA. Carol is then faced with a choice of whether to revise her declared (but false) valuations. If she chooses not to revise, then both players get essentially half of G_1 and half of G_2 for about 50 (honest) points each. However, if Carol revises (essentially optimally) by declaring her true valuations, then Bob will get slightly more than 52 points (as will Carol). Thus, Carol gains 2 points by revising, and, assuming she does revise, Bob gains 2 points by having invoked PA. Ultimately, Carol's deception (and its discovery by Bob) could cost her about 17 points, reducing her from 69 points under AW to 52 points under PA.

Of course, Carol might receive sufficient utility from not confessing to having falsely represented her valuations so that she would be willing to take the 50 points (from sticking with her announced valuations) instead of the 52 points she receives by revising. Notice also that while PA would give Bob only 50.0 points (as opposed to 50.5 points under AW), Bob may regard this sacrifice as minuscule compared with the satisfaction he derives from preventing Carol's exploitation of the situation by reducing her by about 17 points under PA.

We believe this kind of exploitation would be extremely rare if PA were a default option for AW, which we will refer to as the *combined procedure*. Consequently, is this more complex procedure justified? And if PA is invoked, will it be feasible to divide every issue in proportion to the points each player places on it? We think the very conditions that would allow for such exploitation under AW – advance knowledge by one player of the other player's announcements, but not vice-versa – seem highly unrealistic, spies notwithstanding.

A case can be made for changing the combined procedure so that players are not allowed to revise their point allocations, given that PA is selected as a default option. In fact, allowing for revisions (as we just suggested) might make the players more willing to try to exploit their opponents, knowing that they will get a "second chance" if PA is chosen as a default option. On the other hand, if PA is invoked by one player and revisions are allowed, then the player who invoked PA might end up doing better than under AW (as we just saw), given that his or her opponent was exploitative under AW.

Thus, a player who thinks he or she is being exploited under AW would probably have a greater incentive to opt for PA if revisions were allowed under the combined procedure. But this incentive on the part of the exploited player, which would presumably help to deter the exploitative player from attempting exploitation, must be weighed against the exploitative player's own greater incentive to be exploitative if he or she has a second chance. It is not clear where the balance of incentives lies – whether allowing or not allowing revisions under the combined procedure would induce greater truthfulness initially and, consequently, the use of AW. Clearly, a more refined analysis than we have attempted so far is needed to clarify the properties of both versions of the combined procedure.

The winner-take-all feature of AW, used by itself, has an important advantage over proportional sharing under PA: it is applicable to indivisible goods (or issues), except on the one good on which an equitability adjustment may have to be made. Thus, if one is dividing up an inheritance between two heirs, and there are several indivisible goods like a house or a work of art, then AW can more readily be applied than PA – provided, of course, that the equitability adjustment is not made on one of these indivisible goods.

In fact, under AW the players could postpone determining what winning a certain percentage of a good means until the good on which the equitability adjustment must be made is revealed. If, initially, the players are not told who won the larger percentage of that good, but only what this percentage (say, 60 percent) is, then they need only agree on what "winning" with 60 percent (and "losing" with 40 percent) for each player signifies.

Thus, two people might agree that a 60/40 allocation of a late-model car involves the 60-percent person's having sole use of the vehicle for the first three years, after which ownership would be transferred to the 40-percent person. In this example, the people decide on a period of time (three years) for which initial ownership corresponds to the "winning" percentage (60 percent), but other sharing arrangements would certainly be possible.

Since either player could be the 60-percent player, presumably they both would be motivated to be impartial in translating winning and losing into a 60:40 breakdown that could go either way. Alternatively, if the good on which the equitability adjustment is to be made is truly indivisible, the players could arrange an equivalent 60:40 trade on a more divisible good or goods worth about the same amount to both.

4.6 Three or more players

Our analysis of AW, PA, and the combined procedure until now has been quite abstract, illustrated only by hypothetical examples involving Bob and Carol. Although we will discuss in detail possible applications of AW in chapter 5, it is worth noting here that decisions about how to divide things, from cake to countries, are not only ubiquitous but also are usually made without invoking any formal procedures.

Indeed, we introduced this book with a quotation of Harold Nicolson, who indicated how appalled he was at Allied leaders who redrew maps after World War I "as if they were dividing a cake." It was only a generation later that new leaders made the same kinds of decisions as World War II drew to a close:

As Churchill looked around the table at the opening session of the Tolstoy Conference in the Kremlin, at 10 p.m. on October 9, 1944, he decided the moment seemed "apt for business." "Let us settle our affairs in the Balkans," he began, in a phrase much quoted ever after. Then the prime minister wrote out his proposed arrangement on a sheet of plain paper. Russia should have 90 percent predominance in Rumania, Great Britain 90 percent in Greece. They would share fifty–fifty in Yugoslavia and Hungary, and Russia would have 75 percent predominance in Bulgaria. He pushed the paper across to Marshall Stalin who took a blue pencil, made a large tick upon it, and pushed it back.

"After this there was a long silence," recalled Churchill. Finally, the prime minister spoke: "Might it not be thought rather cynical if it seemed we had disposed of these issues, so fateful to millions of people, in such an offhand manner? Let us burn the paper." "No, you keep it," replied Stalin (Gardner, 1993, p. 198).

The next day this kind of talk continued between Anthony Eden, the British foreign minister, and V. M. Molotov, the Soviet foreign minister:

Molotov opened his conversation with Eden by stating that the fifty–fifty ratio proposed for Hungary was unacceptable. The Soviets wanted 75 percent . . . . He argued that Russia must have 90 percent influence . . . [in Bulgaria], as in Rumania. There followed in rapid succession a series of proposals, with Molotov at times offering to trade various percentages in Yugoslavia for near absolute control in Bulgaria and almost the same in Hungary. At one point he attempted to define what these numbers would mean. In Yugoslavia, said the Russian foreign minister, 60/40 meant that Britain would control the coast and Russia the center. Eden eventually agreed to what he thought was a decent compromise – a 20 percent share for Britain in Bulgaria and Hungary, reflected in a two-stage arrangement whereby after the war ended, Russia would allow an allied control commission to function. For that, Molotov agree to equal responsibilities in Yugoslavia. In all, the parceling out of the Balkans was at least reminiscent of the treatment of the Ottoman Empire after World War I, except that the stakes were people rather than oil (Gardner, 1993, p. 202).

These calculations mirror Churchill's recollections in 1956, which contrast sharply with his more lofty principles (section 2.5).

Now, fifty years after the end of World War II, the Balkans are again wracked by virulent conflict, and a new parceling out of land seems likely to occur. The current struggle, however, is less among the great powers for control of the region and more among the local parties, which have been especially gruesome in the former Yugoslavia.

Roiled by long-standing ethnic and religious divisions, Bosnians, Croats, and Serbs have fought a bitter battle for land, and the ridding of opposition groups under their control, sometimes resorting to genocidal policies euphemistically called "ethnic cleansing." While outside parties, primarily under the auspices of the United Nations, have intervened to try to stabilize the situation, their success in preventing fighting at this writing (July 1995) has been extremely limited.

Although alliances among two of the three major ethnic groups in the Yugoslav conflict have been occasionally struck – in an apparent attempt to gain an advantage over the third – these alliances have been fragile and are constantly shifting. At root, the conflict involves at least three players (the Serbian-dominated Yugoslav government in Belgrade has also played a role). Insofar as this conflict and others like it (e.g., that between Israelis and Arabs) cannot be reduced to a two-person game, we are led to ask whether our point-allocation procedures can be extended to larger games.

When there are more than two players, it is not difficult to show that there may be no way of assigning points to players that satisfies efficiency, equitability, and envy-freeness. To illustrate, consider the following example (Riejnierse, 1994):

	G_1	G_2	G_3	Total
A's announced values	40	50	10	100
B's announced values	30	40	30	100
C's announced values	30	30	40	100

The only efficient and equitable allocation turns out to be to give G_1 to A, G_2 to B, and G_3 to C. Obviously, this 40–40–40 allocation is equitable; we show that its efficiency follows from results given in the appendix to this chapter.[16]

But it is not envy-free, because A will envy B for getting G_2, which A considers to be worth 50 points. If we gave G_1 to B and G_2 to A, this allocation would also be efficient, but it would be neither equitable (because each player would get a different number of his or her points) nor envy-free (because B would envy A). Indeed, giving all the goods to one player is efficient, but it fails miserably on the other two properties.

When there are three or more players, it turns out that it is always possible to find an allocation that satisfies two of the three properties: an algorithm that gives both efficiency and envy-freeness has been obtained by Reijnierse and

[16] Varian (1974) showed that in any efficient allocation, there is someone that no one envies (A and C in our example) and someone that envies no one (B and C in our example); for a review of this and related results, see Thomson and Varian (1985).

Potters (1994); linear programs that give both efficiency and equitability have been obtained by Fink and Willson (Fink, 1994; see Willson, 1995, for details); and an equal division of each good to the players gives both equitability and envy-freeness.[17]

It is not clear *a priori* which pair of properties constitutes the most desirable set, and hence which would be the easiest property to give up, if one must sacrifice one property. To the degree that the three major parties to the Yugoslav conflict consider themselves equal players, equitability might be the one most worth preserving. Given an equitable division of the land, we think envy-freeness might be more important to the parties than efficiency, because it would undercut any charges that another player got a "better deal." Thus, the 40–40–40 allocation in the hypothetical example in the text, which is not envy-free, might be worse than an envy-free and equitable allocation that is inefficient.

Of course, this example does not preclude the possibility of all three properties' being satisfied in any particular situation; it says only that it is not always possible to guarantee their satisfaction when there are more than two players. Thus, if three players each like three different goods, an efficient, envy-free, and equitable allocation is obviously possible. Computer simulation could be used to estimate how often a tradeoff among the three properties will be necessary, and under what conditions, but at this point we know little about when, or the degree to which, all three properties can be satisfied in larger fair-division games.

4.7 Conclusions

AW and PA are two practical fair-division procedures that are envy-free and equitable, but only AW guarantees efficiency (at least given truthful revelation). Yet AW's winner-take-all feature makes it potentially vulnerable to

[17] Curiously, PA does not guarantee either equitability or envy-freeness if there are more than two players (even with two players, it will be recalled, PA does not ensure efficiency). To illustrate the failure of PA to satisfy any property in larger games, consider the following three-person, two-good example:

	G_1	G_2	Total
A's announced values	50	50	100
B's announced values	60	40	100
C's announced values	90	10	100

Under PA, A gets 37.5 points, B gets 34 points, and C gets 36 points, which is inequitable. It is inefficient as well, because giving A the fraction $30/37 \cong 0.811$ of G_2, B the fraction $7/30 \cong 0.189$ of G_2 and the fraction $61/111 \cong 0.550$ of G_1, and C the fraction $50/111 \cong 0.450$ of G_1 yields each player 40.6 points, which is efficient as well as being equitable (it is not envy-free: C envies B). Finally, PA is not envy-free, because B would prefer A's $50/200 = 1/4$ allocation of G_1 and $50/100 = 1/2$ allocation of G_2, giving B 35 rather than 34 points.

strategic misrepresentation should one player have information about, or be able to predict, the announced point assignments of the other. PA, because it is much less vulnerable, might therefore be useful to include as a default option should either player think that he or she has been exploited under AW by an opponent who misrepresents his or her preferences. We called the combination of AW and PA, with PA used only as a default option, the combined procedure.

Under this procedure, we assumed initially that the players could revise their allocations if PA were used. But then we went on to consider the default issue when players would be forced to stick with their original announcements. In this case, the player who chose the default option – assuming only one player did – would, we presume, do so primarily to punish his or her opponent for trying to be exploitative.

On first blush, this calculation would seem to be irrational. But it is not if the punisher attributes sufficiently great utility to ensuring a truly equitable solution, based on truthfulness. That is, by hurting himself or herself – perhaps only slightly, as we illustrated by the example in section 4.5 – in order to punish his or her opponent severely for making a false announcement, the punisher may in fact derive a net benefit from choosing PA: letting one's opponent go unpunished under AW for his or her false announcement may be more painful than suffering a slight loss under PA.[18]

Whether the combined procedure permits or does not permit the players to revise their point allocations, however, we think that the choice of PA as a default option would rarely occur. It seems highly unlikely that a player would be able to ascertain that he or she would benefit by exercising the default option, especially on negotiations like those in which – absent spies – it seems virtually impossible to anticipate an opponent's point assignments with any precision on several different issues.[19]

Used alone, AW has a major advantage over the combined procedure (in which PA may be used): issues may be indivisible, except on the one issue on

[18] Nuclear deterrence, interestingly enough, has a similar rationale. Suppose a country threatens to escalate a conflict to nuclear war, which could result in horrendous damage to itself as well to its opponent if the threat were carried out. Just as this seemingly irrational threat is rational if it is successful in deterring a conflict that could escalate to nuclear war (Brams, 1985b; Zagare, 1987; Brams and Kilgour, 1988; Powell, 1989), the possibility that PA might be used – without allowing players to revise their allocations – under the combined procedure is rational if it deters a player from trying to be exploitative.

[19] Besides the Panama Canal treaty negotiations (to be discussed in chapter 5), Raiffa (1982, chapter 10) gives a hypothetical example of negotiations involving ten issues in which he suggests that players, "thrashing around" (p. 139), might eventually find the so-called Pareto-optimal frontier (i.e., where all allocations are efficient, but only one is equitable). But he also says (p. 288) that some "systems or mechanisms for conflict resolution . . . are far better than unstructured improvisation" (p. 288). We agree, offering AW as our candidate; for a formalization of Raiffa's approach, see Ehtamo, Verkamma, and Hämäläinen (1994).

which an equitability adjustment must be made, because a player wins or loses completely on each. Although the players need to spell out beforehand what each side obtains when it wins or loses under AW, only on the issue on which an equitability adjustment must be made will a finer breakdown be necessary.

This breakdown, however, can await the application of AW. As we suggested in section 4.5, the players could be told the split on the equitability-adjustment issue but not which player was the relative winner or loser. Not knowing whether they won or lost the larger percentage on this issue, the players would be motivated to reach a fair-minded agreement on how winning by this percentage (by either player) will be translated into a settlement.

Some of the most severe conflicts in the world are essentially two-person conflicts, such as those between Hindus and Muslims in India and Hutus and Tutsis in Rwanda – not to mention the persistent, if now diminished, struggles between Catholics and Protestants in Northern Ireland and blacks and whites in South Africa. But other conflicts, such as that in the former Yugoslavia which was briefly discussed in section 4.6, have more than two parties. At an international level, the Israeli-Arab conflict involves several different countries, as well as factions within some of the countries (including Israel), that make it decidedly an *n*-person game.

As we showed, there is no allocation procedure that can guarantee the three properties of efficiency, envy-freeness, and equitability when there are more than two players. The fact that AW guarantees all three in the two-person case is encouraging, despite its theoretical (but probably not practical) vulnerability to manipulation. In chapter 5 we turn to applications of AW to some two-person conflicts – the Panama Canal treaty negotiations in the 1970s and two divorce cases.

Appendix

Theorem 4.1 AW produces an allocation of the goods, based on the announced valuations, that is:

(1) efficient: *any allocation that is strictly better for one player is strictly worse for the other;*
(2) equitable: *player I's announced valuation of his allocation is the same as player II's announced valuation of her allocation;*
(3) envy-free: *neither player would trade his or her allocation for that of the other.*

Proof. To establish the efficiency of AW, we first prove three claims:

Claim 1. Suppose we have an allocation wherein:
(i) Player I values G_i at least as much as player II does;
(ii) Player II values G_j at least as much as player I does;
(iii) Player I possesses the amount $S \subset G_i$;
(iv) Player II possesses the amount $T \subset G_j$.
Then if a trade of S for T yields an allocation that is better for one player, it is worse for the other.

Proof. Assume that a and a' are player I's values of G_i and G_j, respectively, and that b and b' are player II's values of G_i and G_j, respectively. Thus, $a \geq b$ and $a' \leq b'$. Let $|S|$ denote the fraction of G_i that S is, and let $|T|$ denote the fraction of G_j that T is. Assume that the trade strictly benefits player I. Then

[what player I gets] > [what player I has],

and so

$$|T|(a') > |S|(a).$$

But because $b' \geq a'$ and $a \geq b$, we have

$$|T|(b') > |S|(b),$$

and so

[what player II has] > [what player II gets].

Hence, the trade is strictly worse for player II. Q.E.D.

Claim 2. Suppose we have an allocation wherein:
(i) Player I's values, G_i and G_j, are a and a', respectively;
(ii) Player II's values, G_i and G_j, are b and b', respectively;
(iii) Player I possesses the amount $S \subset G_i$;
(iv) Player II possesses the amount $T \subset G_j$;
(v) $a'/b' \leq a/b$.
Then if a trade of S for T yields an allocation that is better for one player, it is worse for the other.

Proof. Suppose, for contradiction, that a trade of S for T yields an allocation that is strictly better for one player and no worse for the other. Then we have

[what player I gets] ≥ [what player I has]

and

[what player II gets] ≥ [what player II has],

with at least one of the inequalities strict. Thus

$$|T|(a') \geq |S|(a),$$

and

$$|S|(b) \geq |T|(b'),$$

with at least one of these inequalities strict. Multiplying the first inequality by b on both sides, and the second by a on both sides, yields

$$|T|(a')(b) \geq |S|(a)(b),$$

and

$$|S|(a)(b) \geq |T|(a)(b').$$

Hence

$$|T|(a')(b) \geq |T|(a)(b'),$$

and so

$$(a')(b) \geq (a)(b').$$

Moreover, this inequality is strict since one of our first two inequalities was. Consequently, $a'/b' > a/b$, in contrast to (v) in claim 2. Q.E.D.

Claim 3. If a given allocation is not efficient, then there are goods G_i and G_j, and sets $S \subset G_i$ and $T \subset G_j$, such that a trade of S for T yields an allocation that dominates the given one.

Proof. Since the given allocation is not efficient, we can choose disjoint sets S' and T' so that player I possesses S', player II possesses T', and a trade of S' for T' is better for one (say, player I) and no worse for the other. The set S', however, may not be a subset of a single good – as we want (and will now show how to obtain) – but it certainly can be written as the disjoint union $S_1 \cup \ldots \cup S_m$ of sets that *are* subsets of single goods. Player II can now split T' into disjoint sets $T_1 \cup \ldots \cup T_m$ (which are not necessarily subsets of single goods) so that a trade of T_i for S_i yields an allocation that is no worse for him or her than the given allocation.[20]

Because of efficiency, there must now exist at least one i so that player I finds the allocation resulting from a trade of S_i for T_i strictly preferable to the given allocation. (This also uses weak additivity of preferences.) Recall that S_i

[20] We assume here weak additivity of preferences. That is, if A and C are disjoint sets, and if a player thinks A is at least as large as B, and C is at least as large as D, then that player thinks A ∪ C is at least as large as B ∪ D.

is a subset of a single good, but T_i may not be. Nevertheless, we can write T_i as the disjoint union of sets which are subsets of single goods and then proceed, as before, to obtain first a set $S \subset S_i$ and then a set $T \subset T_i$ so that a trade of S for T is strictly better for player I and no worse for player II. Q.E.D.

The theorem now follows easily from the three claims. Suppose that the allocation from the procedure is not efficient, and choose sets $S \subset G_i$ and $T \subset G_j$ as guaranteed to exist by claim 3 when there is not an efficient allocation. Assume player I had the advantage in the winner-take-all part of the procedure, so any transference of goods in the second step of the procedure was from player I to player II. Since player I possesses S, he or she values good G_i at least as much as player II does (say, $a \geq b$). It now follows from claim 1 that player I values good G_j strictly more than player II does (say, $a' > b'$). Since player II possesses T, he or she must have received it from player I in the transfer stage of AW. However, player I still possessed part of G_i, so all of it was not transferred to player II. Thus, we must have $j < i$ and so $a'/b' \leq a/b$. This last statement contradicts claim 2 and thus completes the proof of efficiency.

Equitability is clear by construction, and envy-freeness follows from efficiency and equitability. That is, if the allocation were not envy-free, then both players would receive fewer than 50 points and, hence, equal division would contradict efficiency.[21] Q.E.D.

In section 4.6 we claimed that the only efficient and equitable allocation in the following three-person, three-good example is to give all of G_1 to A, all of G_2 to B, and all of G_3 to C:

	G_1	G_2	G_3	Total
A's announced values	40	50	10	100
B's announced values	30	40	30	100
C's announced values	30	30	40	100

We will now verify this claim, based on the following development of a necessary condition for efficiency.

Suppose an allocation of *m* homogeneous goods among *n* players is efficient (in a context where utilities are reflected by point assignments, as in this chapter). Choose any two players (call them player I and player II) and any two goods (call them G_i and G_j). Assume the point assignments are as follows:

[21] It is worth noting that if a two-person point-allocation procedure is equitable, then either the allocation is envy-free or a trade between the two players would yield an envy-free allocation. Thus, we never have a situation wherein one player wants to trade and the other does not when an allocation is equitable.

	G_i	G_j
Player I	a	a'
Player II	b	b'

We compare the products obtained by multiplying the entries on each of the two diagonals. Thus, looking at $a \times b'$ and $a' \times b$, we seek the diagonal corresponding to the smaller product (assuming the two are not equal), which we call the *light diagonal*. Assume here it is ab', which we can picture as follows:

	G_i	G_j
Player I	a	
Player II		b'

If player I possesses the non-zero amount S of G_i, and player II possesses the non-zero amount T of G_j, then we say that the "light diagonal is shared." With this terminology, we now state

The light-diagonal condition

If an allocation is efficient, then no two players ever share a light diagonal.

Proof. Assume that players I and II share the light diagonal ab' we just illustrated. Have player I choose $S' \subset S$ from the part of G_i allocated to him or her and also $T' \subset T$ from the part of G_j allocated to player II so that player I values S' and T' the same. Thus, if a trade is made of T' for S', then the allocation is no worse (in fact, it is exactly the same) for player I. We claim that this trade, however, yields an allocation that is strictly better for player II.

Let $|S'|$ denote the fraction of G_i that S' is, and let $|T'|$ denote the fraction of G_j that T' is. Because player I values S' and T' the same, we have:

$$|S'| \times a = |T'| \times a'. \tag{4.1}$$

Our assumption that $a \times b' < a' \times b$ can also be written

$$a'/a > b'/b. \tag{4.2}$$

Player II is giving up $|T'| \times b'$ and gaining $|S'| \times b$. Thus, we want to show that

$$|S'| \times b > |T'| \times b'. \tag{4.3}$$

But (4.1) and (4.2) together yield (4.3) as follows:

$$|S'| \times b = |S'| \times a \times (b/a) = |T'| \times a' \times (b/a) \quad \text{from (4.1)}$$
$$= |T'| \times (a'/a) \times b > |T'| \times (b'/b) \times b = |T'| \times b' \quad \text{from (4.2)}$$

as desired.

Returning to our three-person, three-good example, suppose the 40–40–40 allocation is not efficient. Then there exists some other allocation in which each player does strictly better.[22] We can assume, without loss of generality, that this other allocation is itself efficient.[23]

Claim 1. Player B receives none of G_1.

Proof. Suppose player B receives some of G_1. Then player A receives none of G_2 (or else A and B would share the light diagonal 30 x 50 < 40 x 40), and player A receives none of G_3 (or else A and B would share the light diagonal 30 x 10 < 40 x 30). Thus, player A receives only G_1 and hence fewer than 40 points, since we are assuming player B has also received some of G_1. Thus, the allocation is worse for player A, contrary to our assumption. Q.E.D.

Claim 2. Player B receives none of G_3.

Proof. Suppose player B receives some of G_3. Then player C receives none of G_1 (or else B and C share the light diagonal 30 x 30 < 30 x 40), and player C receives none of G_2 (or else B and C share the light diagonal 30 x 30 < 40 x 40). Hence, player C receives only G_3, and thus fewer than 40 points, because we are assuming player B also received some of G_3. Again, this is a contradiction. Q.E.D.

It now follows from claims 1 and 2 that player B receives only G_2 and thus at most 40 points. This contradiction completes the proof. Q.E.D.

Notice that the claim 2 in the proof of Theorem 4.1 shows that if a trade of S' for T' is strictly better for one player and no worse for the other, then $a \times b'$ < $a' \times b$. Hence, it is the light diagonal.

[22] Recall our general explanation for why this is the case in note 14 in section 2.8. In the present example, if the 40–40–40 allocation is not efficient, then there is some allocation in which at least one of the players does strictly better and the other two do no worse. But then the player who did strictly better can give a little of his or her allocation to each of the other players so that they, too, do strictly better.

[23] We lose no generality by the following argument: If $x_0 \geq y_0 \geq z_0 > 40$, and we have an "$x_0 - y_0 - z_0$ allocation," then

$$X = \{(x,y,z): x \geq x_0, y \geq y_0, z \geq z_0, \text{ and there exists an } x-y-z \text{ allocation}\}$$

is a compact set by a well-known theorem of Lyapounov (1940). The function $f(x,y,z) = x + y + z$ is a continuous function on X and thus achieves a maximum and minimum. It is easy to see that this maximum is an efficient allocation.

Theorem 4.2 Assume there are two goods, G_1 and G_2, all true and announced valuations are restricted to the integers, and suppose player I's announced valuation of G_1 is x, where $x \geq 50$. Assume player II's true valuation of G_1 is b. Then player II's optimal announced valuation of G_1 is:

$$\begin{array}{ll} x+1 & \text{if } b > x \\ x & \text{if } b = x \\ x-1 & \text{if } b < x. \end{array}$$

Proof. The proof requires the following five claims:

Claim 1. Suppose player I's announced valuations of G_1 and G_2 are x and x', respectively, and player II's announced valuations are the truthful assignments b and b', respectively. Assume x is the largest of the four valuations. Then player I receives the fraction $100/(x + b)$ of G_1 and none of G_2, whereas player II receives the rest of G_1 plus all of G_2, which are allocations that are equal in terms of each player's valuations.

Proof. Let α denote the fraction of G_1 that will be retained by player I. The equitability adjustment requires that

$$x\,\alpha = b\,(1 - \alpha) + b',$$

so

$$\alpha = (b + b')/(x + b),$$

where $b + b' = 100$. Q.E.D.

Claim 2. For player II, the announced value of $x + 1$ for G_1 is optimal among all announcements $y > x$.

Proof. An announcement by player II of $y > x$ makes y larger than any of the four values posited in claim 1. The equitability adjustment now requires a choice of α, as in the proof of claim 1, such that

$$y\,\alpha = x\,(1 - \alpha) + x',$$

so

$$\alpha = 100/(x + y).$$

Because player II's true valuation of G_1 is b, she retains only the fraction α of it, or $100b/(x + y)$ points. This is largest when y is smallest. Q.E.D.

Claim 3. For player II, the announced value of $x - 1$ for G_1 is optimal among all announcements $y < x$.

Proof. If player II announces a valuation of $y < x$ for G_1, then his or her payoff depends on whether or not $100 - y > x$. If $100 - y > x$, he or she receives none of G_1 and a fraction of G_2, which gives player II a payoff of

$$(100 - b)[100/(100 - y + 100 - x)] = (100 - b)[100/(200 - x - y)].$$

If $100 - y \leq x$, then player II receives a payoff of

$$100 - b + b[1 - 100/(x + y)] = 100[1 - b/(x + y)].$$

Either way, player II's payoff is maximized by choosing y as large as possible, which is $x - 1$ for $y < x$. Q.E.D.

Claim 4. In the case where $x = 50$ and $b \neq 50$, player II's payoff from an announcement of $x - 1$ yields a higher payoff for player II than does an announcement of $x + 1$ if and only if $b < x$.

Proof. Straightforward.

Claim 5. If $x > 50$, then an announcement of $x - 1$ yields a higher payoff for player II than does an announcement of $x + 1$ if and only if $b < x$.

Proof. Player II's payoff from an announcement of $x - 1$ is

$$(100 - b) + b\,[1 - 100/(2x - 1)],$$

whereas his or her payoff from an announcement of $x + 1$ is

$$100b/(2x + 1).$$

The former payoff is greater than the latter if and only if

$$100[1 - b/(2x - 1)] > 100b/(2x + 1).$$

Simplifying this inequality yields the following sequence of inequalities:

$$1 - b/(2x - 1) > b/(2x + 1),$$
$$(2x - 1 - b)/(2x - 1) > b/(2x + 1),$$
$$4x^2 - 2x - 2xb + 2x - 1 - b > 2xb - b,$$
$$4x^2 - 4xb > 1,$$
$$b < x - 1/4x,$$
$$b < x.$$

Q.E.D.

Claims 1–5, together with the trivial observation that truthfulness is the best policy if player II's true valuations coincide with player I's announced valuations, complete the proof of the theorem. Q.E.D.

Theorem 4.3. PA produces an allocation of the goods, based on the announced valuations, that is equitable and envy-free.

Proof. We first show that PA is equitable. The payoffs, $P_I(\vec{x},\vec{y})$ and $P_{II}(\vec{x},\vec{y})$, of players I and II are the weighted sums of their point assignments to each good multiplied by their fractional allocations:

$$P_I(\vec{x},\vec{y}) = x_1^2/(x_1+y_1) + \cdots + x_k^2/(x_k+y_k)$$

$$P_{II}(\vec{x},\vec{y}) = y_1^2/(x_1+y_1) + \cdots + y_k^2/(x_k+y_k).$$

Because we assume that the players may not be truthful, these payoffs are only apparent, as we noted for the payoffs under AW as well.

To see that $P_I(\vec{x},\vec{y}) = P_{II}(\vec{x},\vec{y})$, and PA is therefore equitable, notice that

$$P_I(\vec{x},\vec{y}) - P_{II}(\vec{x},\vec{y}) =$$

$$\sum_{i=1}^{n} \frac{x_i^2 - y_i^2}{x_i + y_i} = \sum_{i=1}^{n} \frac{(x_i - y_i)(x_i + y_i)}{x_i + y_i} = \sum_{i=1}^{n} (x_i - y_i) = 0.$$

To show that PA is envy-free, we must show that if $0 \le x_i \le 1$, $0 \le y_i \le 1$, and

$$\sum_{i=1}^{k} x_i = \sum_{i=1}^{k} y_i,$$

then

$$\sum_{i=1}^{k} x_i^2/(x_i + y_i) \ge 1/2 \text{ and } \sum_{i=1}^{k} y_i^2/(x_i + y_i) \ge 1/2.$$

(That is, the payoffs to player I and to player II are at least one-half the 100 points.) Because of the equitability of PA (just demonstrated), it suffices to show that:

$$\sum_{i=1}^{k} (x_i^2 + y_i^2)/(x_i + y_i) \ge 1. \tag{4.4}$$

The following demonstration, due to Julius B. Barbanel, improves upon what we had in the earlier version of this proof. Since

$$\sum_{i=1}^{k} (x_i + y_i)/2 = (1/2)(\sum_{i=1}^{k} x_i + \sum_{i=1}^{k} x_i) = (1/2)(1 + 1) = 1,$$

it suffices to show that, for each i,

$$(x_i^2 + y_i^2)/(x_i + y_i) \geq (x_i + y_i)/2.$$

Suppose this inequality fails. Then

$$2x_i^2 + 2y_i^2 < x_i^2 + 2x_iy_i + y_i^2.$$

But then

$$x_i^2 - 2x_iy_i + y_i^2 < 0,$$

so $(x_i - y_i)^2 < 0$. This is a contradiction. Thus, inequality (4.4) is satisfied, so PA is envy-free. Q.E.D.

5 Applications of the point-allocation procedures

5.1 Introduction

Our thesis in this chapter is that AW is not simply a fair-division procedure with desirable theoretical properties. It is also one that is eminently practicable in a variety of situations, ranging from international disputes to divorce settlements.

We begin section 5.2 by showing how AW might have been used in facilitating the resolution of the dispute between the United States and Panama over the Panama Canal treaty, which was negotiated in the 1970s and involved ten issues. In sections 5.3 and 5.4 we illustrate the application of AW to two divorce cases, one hypothetical and one real (from New York state). In the real case, AW probably would have produced a different resolution, which the dissenting judge in this case, when heard on appeal, supported. As a final application of AW, we revisit in section 5.5 the problem of dividing up the Vermont estate, which was first discussed in section 1.2, and compare this division with the division obtained using Knaster's procedure, which was described in section 3.2.

Knaster's procedure, when applied to two-party disputes involving multiple issues, bears some resemblance to AW insofar as the bids that the players make may be thought of as their placing points on the items to be divided. But there is a major difference, even in the case when the bids of the players sum to the same total amounts: Knaster's procedure requires that players have bankrolls, which they can draw upon to make side payments to other players, whereas AW requires a possible redistribution of some of the goods themselves. We shall illustrate this and other differences and indicate why we believe AW to be a superior procedure in most real-world applications.

5.2 The Panama Canal treaty negotiations

If the goods are issues over which there is negotiation, there is often a mix, with some issues more divisible than others.[1] Consider, as a case in point, the

[1] This section is adapted from Brams and Taylor (1994b).

Table 5.1. *Point allocations in Panama Canal treaty negotiations*

Issue	United States	Panama
1 US defense rights	22	9
2 Use rights	22	15
3 Land and water	15	15
4 Expansion rights	14	3
5 Duration	11	15
6 Expansion routes	6	5
7 Compensation	4	11
8 Jurisdiction	2	7
9 US military rights	2	7
10 Defense role of Panama	2	13
Total	100	100

Source: Raiffa (1982, Table 10, p. 177)

Panama Canal treaty negotiations, in which the United States and Panama agreed in June 1974, after two rounds of negotiations, on a definition of the ten major issues shown in table 5.1. Raiffa (1982, pp. 176–7) reports that a consulting firm then interviewed members of the US negotiating team, headed by Ambassador-at-Large Ellsworth Bunker, not only about the importance that the United States attached to these issues but also the importance, as viewed by the Americans, that Panama attached to the same issues.

These assessments are shown in table 5.1. Under AW, the United States wins on issues 1, 2, 4, and 6 (its points are underscored on these four issues in table 5.1), giving it 64 points, whereas Panama wins on issues 5, 7, 8, 9, and 10 (its points are underscored on these five issues), giving it 53 points. The players tie with 15 points each on issue 3, which we give to the player (the United States) with the larger total on untied issues (section 4.3). This gives the United States an initial allocation of 79, and Panama an initial allocation of 53, points.

Issue 3 has the smallest point ratio (15/15 = 1.0) and becomes, therefore, the first issue used in the equitability adjustment. If we let α denote the fraction of that issue that the United States will retain, then, because the United States has 64 points without issue 3 and Panama has 53 points, α must satisfy the following equation:

$$64 + 15\alpha = 53 + 15(1 - \alpha).$$

Solving for α, we find that $\alpha = 2/15$. Thus, the United States should retain 13.3 percent of its position (2 points), and Panama 86.7 percent of its position (13 points) on issue 3 to equalize their point totals. This results in both the

United States (64 + 2 = 66) and Panama (53 + 13 = 66) receiving an envy-free and equitable assignment of exactly 66 points under AW.

The 66 points that each of the players receives under AW compares with 57.9 points that PA would give them. Thus, AW provides a 14 percent efficiency improvement and is, presumably, the choice the players would make in this example. Moreover, issue 3, "land and water," is one of the more divisible issues and so, presumably, one on which the players could agree on a "resolution level" (Keeney and Raiffa, 1991, p. 132) that would give Panama about 7/8 of its way.

In fact, the only issue for which Raiffa (1982, pp. 176–7) – using slightly different terminology – indicates there is not a divisible "bargaining range" is issue 6, "expansion routes," for which there were only three possible choices. Although most issues in negotiations like these are probably more or less divisible, it is important, especially if the combined procedure (and, therefore, possibly PA) is used, that the players agree beforehand on resolution levels in order to prevent quarreling on what so-and-so's winning x percent on that issue means. On the other hand, as we suggested in section 4.5, under AW this designation could be postponed until the one issue on which an equitability adjustment must be made is known.

Perhaps a greater problem in applying any of the procedures is the assumption of each that the points across all issues are additive (section 4.3), though Keeney and Raiffa (1991) offer practical suggestions for helping to ensure this to be the case. The technical validity of this assumption depends on the issues' being *separable* – that is, that the amount that a player wins on one issue does not depend on his or her winning on other issues, so the issues can be treated independently (we defined the separability of goods in section 4.3).

A possible solution to this problem would be to allow the players to lump nonseparable issues together, such as issue 1 (US defense rights) and issue 9 (US military rights) for the United States. But the assumed linearity of points is also a potential problem. If the United States wins a particular amount on these two issues when lumped together (say, 60 percent), it might make more sense for it to give up more than 40 percent on one issue and less than 40 percent on the other than be forced under PA – or AW, if there is an equitability adjustment on this combined issue – to give up prescribed, but non-complementary, amounts on each of these two rights issues.

An extreme form of lumping would be to have one player divide the issues into two "packages," each of which he or she values equally. For example, the United States values the sets $\{1, 2, 6\}$ and $\{3, 4, 5, 7, 8, 9, 10\}$ at 50 points each, whereas Panama values the first set at 29 points and the second set at 71 points. Even if Panama knew this was a 50–50 division for the United States, its 29–71 truthful response would be near optimal under PA (Panama's optimal response to 50–50 would be 28–72).

It turns out that truthfulness under PA would give each side only 52.3 points for these two packages, which is considerably below the 57.9 points that the players would receive by making truthful assignments to the ten different issues. A similar degradation occurs under AW, in which the players receive 58.7 points when there are only two packages but 66.0 points when there are ten issues.

There is not only a cost to lumping, but it also increases as the two packages approach equality for the players. For example, the packages $\{1, 2, 8, 9, 10\}$ and $\{3, 4, 5, 6, 7\}$ are 50–50 for the United States and 49–51 for Panama, which yields the players only 50.5 points under AW and 50.0 points under PA. We conclude from this example that two players, in order to maximize their point totals under both AW and PA, would be well advised to apply these procedures to *as many different issues as they can reasonably make separable and additive.*

5.3 A hypothetical divorce settlement

Having illustrated the applicability of AW and the combined procedure (with PA as a default option) to an international dispute, we could turn to any number of other kinds of disputes, such as those that occur between management and labor in company-union bargaining, to illustrate their further application.[2] But it is surely divorces – more than any other kind of dispute – that cry out for better settlement procedures and for which AW, in particular, seems especially well suited.[3]

Divorces may be not only bitter and acrimonious but also very costly to the two parties as lawyers' fees accumulate. Such a simple dispute-resolution procedure as AW may not please some divorce lawyers, but we believe many may be able to use it to their advantage, which we shall say more about in section 5.6.

Although AW would seem well suited for reaching a reasonable settlement of the property in a divorce, a large indivisible item like a house could pose a problem, especially if both sides value it about equally. If the house represents more than 50 percent of the value for both parties, then it is not clear – short of selling it – how AW could be applied.

But divorce agreements may involve not just the allocation of material property but also the future compensation of one party by the other. Such compensation over a sufficiently long period of time, if available, may well balance out the immediate gain that one party receives from obtaining, say, the house. In addition, compromises on matters like custodial or visitation rights,

[2] This and the next three sections are adapted from Brams and Taylor (1994c).

[3] Crawford (1977, p. 241) argues that certain emendations in divide-and-choose, which allow for trading between the players, are also applicable.

if there are children (Elster, 1989; Maccoby and Mnookin, 1992), may be another tradeoff that can be included to facilitate the achievement of equitability under AW.

To illustrate how AW might be used in a divorce settlement, consider the hypothetical case of Tom and Mary presented in *Negotiating to Settlement in Divorce* (1987, pp. 166–9), a manual for lawyers. There are three major issues to be resolved, with the preferences of Tom and Mary on each recounted below (quotations are taken from the case):

1 Custody

Tom: "Tom wants to be seen as vigorously trying to obtain sole custody of [their son] John . . . [so that] no one will ever be able to convince John that his father did not put up a fight for sole custody. In fact, Tom would prefer joint custody, and, if pushed, would admit to being willing not to have custody at all, provided that he could have flexible and extensive visitation rights. . . [He will] concede the question of sole custody *only as a last resort*" [italics in original] (p. 167).

Mary: "John's great grandfather established a trust now worth several million dollars . . . [that will] be distributed to [the] . . . great-grandchildren . . . provided that they have been raised 'in the manner and tradition of the McCrees.' Apart from John, there is only one other great grandchild, Miles, the son of Mary's sister. Mary and her sister have not spoken for years. Their animosity is so great that Mary knows that given the slightest opportunity, her sister will do anything she can to frustrate John's claim to the trust capital. Mary is concerned that if Tom gets custody – even joint custody – of John, Mary's sister will argue that John was not raised [properly]. Accordingly, . . . [Mary wants] to do everything possible to obtain sole custody . . . [and] may not settle [otherwise]" (p. 168).

2 Alimony

Tom: "Tom is adamant [against] . . . his having to pay alimony, because 'she will always be coming back for more.' At very worst, he would accept having to pay a 'small' amount of alimony provided he could be assured that it was not modifiable and would only be payable for a fixed, 'short' period" (p. 167).

Mary: "Mary is furious that Tom wants to leave her just when the money is starting to come in [Tom is about to graduate from law school], after what she sees as all her hard work. Thus Mary's [desire is to] . . . 'take him for every cent [she] can get' " (p. 168).

3 House

Tom: "Tom would like to keep the house. At any rate, he feels he is entitled to $10,000 of the equity, representing his grandfather's gift, and half of any equity beyond that" (p. 167).

Mary: "Mary . . . does want to continue to live in the house" (p. 168).

On the basis of these preferences, Tom and Mary might plausibly make the following point assignments, with the larger of the two assignments underscored:

	Custody (sole)	Alimony	House	Total
Tom	25	<u>60</u>	<u>15</u>	100
Mary	<u>65</u>	25	10	100

These reflect Mary's great interest in obtaining sole custody, Tom's in minimizing if not avoiding altogether paying alimony, and the lesser interest of each in gaining possession of the house.[4]

Under AW, Mary would win sole custody of John, whereas Tom would get his way on alimony and, initially, get the house. However, because Tom ends up with 75 points and Mary only 65 points after the assignment of points to the winners on each issue, Tom must give back points on the house, which has a smaller ratio of Tom's points to Mary's points (15/10 = 1.5) than does alimony (60/25 = 2.4).

Equitability will be achieved when Tom's points and Mary's points are set equal, with Tom getting the fraction α of the house and Mary getting the fraction $1 - \alpha$. Solving

$$60 + 15\alpha = 65 + 10(1 - \alpha)$$

for α gives $\alpha = 3/5$. Thus, Tom is entitled to somewhat more than half the house and Mary somewhat less. In terms of points, Tom will get 60 + 9, and Mary 65 + 4, or 69 points each. This total is greater than that which each party obtained in the Panama Canal treaty dispute (66 points), because in this case the parties have almost opposite preferences on the two major issues, enabling them both to do better in getting the things they want. If PA were used in this case, the players would receive only 57.9 points, a 16 percent decrease over AW.

Curiously, the house in this example is the least important item for both disputants. Whoever retains it will presumably have to compensate the other party for its fractional share. Which person is entitled to the 3/5 share should

[4] We readily admit that the players in this case as well as the real divorce case we discuss next (sections 5.4 and 5.5) could make other plausible numerical point assignments. Indeed, each side might do "sensitivity analysis" to see how its own assignments might affect the outcome for different possible assignments of the other side. In this hypothetical case, since Mary puts a great premium on custody, and Tom on alimony, it is only the house on which they are likely to come out reasonably close, with perhaps Mary ending up putting more points on it than Tom (unlike in our example). But relatively small variations in their point assignments would not alter the fact that the house would remain the item on which an equitability adjustment is made. Hence, we conclude that custody and alimony in this case are insensitive to small point variations, but who gets what share of the house is affected by such variations.

probably be kept secret until both parties are able to hammer out an agreement on what winning this proportion means, in terms of a settlement, when each is either the 3/5 or 2/5 party.[5]

The authors of this hypothetical divorce case, it is worth noting, threw in some interesting twists that we did not mention previously. First, Tom is reported as having an affair with a fellow law student, Kate, whom he has "no wish to marry or live with . . . after his divorce" (p. 166). Nonetheless, he does not want Mary to find out about Kate, which he thinks is "quite possible if the matter is not settled before trial" (p. 166). Moreover, if his affair should become public in the course of the trial, Tom fears that his career as a lawyer would be jeopardized if not ruined.

Since the break-up of their marriage, Mary "has become involved with another man, and she thinks this relationship has the possibility of maturing into something permanent" (pp. 168–9). Moreover, even before Tom and Mary's marriage had broken up, "some of [Mary's] friends thought they saw Tom with his arm around another woman. However, it was some distance away, and the friends were not absolutely sure it was Tom" (p. 169).

This spicy information, if found out by the other party, would probably not alter the point allocations of Tom and Mary that we posited earlier. It is likely, however, to impel them to try to expedite settlement of their case.

What is ironic in this case is Tom's apparent need to exhibit his strong concern for the welfare of his son, John. But in fact Tom was ready to cede sole custody of John to Mary. In actual negotiations, Tom would presumably posture on the custody issue in order to extract as much as possible on the alimony issue.

Such posturing becomes problematic under AW, because if Tom places too many points on custody, he might lose on alimony, which is his big worry. AW, as we see it, would largely eliminate posturing of this sort, because it is "hard" points that count in the end rather than "soft" – and perhaps insincere – concerns one might try to use to advance one's bargaining position.[6]

Is AW like a modern-day King Solomon, at least insofar as it renders

[5] After Tom and Mary submit their point assignments, and the calculation described in the text is made by a neutral third party whom we will call a *referee*, we suggest the following protocol to implement the suggestion we offered in section 4.5: Tom and Mary would be told by the referee that the house is the item on which an equitability adjustment must be made. It will be "split" 3:2 between them – in terms of one party's having to compensate the other for its share – but they would not be told who is entitled to the larger share. Given this protocol, Tom and Mary would, we believe, be motivated to come to an agreement on what being the 3 "winner" and the 2 "loser" mean, because each would want to protect himself or herself in either contingency. If such compensation is infeasible, the house might be sold and the proceeds split 3:2.

[6] To make things easier for Tom in the case at hand, we would add that there is nothing in AW that requires that it be revealed, or made part of the public record, that his son, John, was worth "only" 25 points to him.

false claims – like that of the impostor in the Bible story – counterproductive (section 1.1)? It certainly does not do so by tricking the disputants, as Solomon did, but instead, we would argue, by making truthfulness a generally sound strategy.

5.4 A real divorce settlement

Erlanger, Chambliss, and Melli (1987, p. 583) characterize informal processes used in divorce settlements as "often contentious, adversarial, and beyond the perceived control of one or both parties." AW, we believe, would give the parties some greater measure of control by establishing clear "rules of the game" by which marital property is distributed. But are there any real-life instances in which the discipline of AW might have helped both parties in a divorce?

Although no case can simply be rerun to see what an alternative dispute-resolution procedure would have produced, the following divorce case illustrates how AW can take some of the arbitrariness out of the allocation of property to the spouses, which failed in part to mirror the preferences of both (but especially the wife's). In fact, the dissenting judge in this case, when it was heard upon appeal, offered *prima facie* support for our argument that there was an alternative settlement of this case that would have benefited both parties. It might have been implemented, we believe, if AW had been used.

Before discussing the details of this case, it is worth pointing out that property–division laws vary from state to state. Most now incorporate some form of an "equal division" standard, such as the Marital Equitable Distribution Law enacted by New York state in 1980. Although this law was not designed as a substitute for a judge's weighing of the conflicting claims of the two sides, it does provide a set of guidelines to determine what distribution of property is equitable. Given that certain criteria are met, the court must subject the distribution of property to a 50-50 standard.[7]

The case of *Jolis v. Jolis*, 446 N.Y.S.2d 138, was the first case heard in New York City by the state Supreme Court after the law took effect on July 19, 1980 (the trial commenced on December 5, 1980). Several elements in this case – which was decided on October 30, 1981, after many delays – were cited as warranting "equal division of marital property" (p. 138), including the

[7] This standard has been much criticized; see, for example, Fineman (1991), who, on the basis of "horror stories" and statistical studies, argues for "the abdication of equality" because "I believe women and children will fare better under legal rules that reference their material and emotional circumstances, not grand theoretical abstractions" (p. 11). Elsewhere (p. 204) Fineman notes that in the case we describe next, *Jolis v. Jolis*, "the equity model is a form of property reallocation or asset-sharing based on other women's social or cultural disadvantages and not on their own circumstances."

marriage's duration of 41 years (the couple spent 33 of these years together), the wife's relinquishment of her early and successful career, her role in the marriage and the raising of the couple's four sons, her "spousal companionship," and her "lack of skills or training which would enable her to undertake a self-supporting career" (p. 150).

There were both real estate and liquid assets to be divided in this case. But in the absence of any formal procedure to divide the property, a lengthy dispute ensued, which eventually reached the appellate division of the Supreme Court. On appeal the five-person Court did not reach a consensus on a fair division of the property, with one of the justices writing a strong dissent of the majority decision.

This dispute was very costly in time and money to the couple, who had enjoyed a life of substantial wealth, luxury, and gracious living. The wife had given up a budding career as a successful singer when she married, with the Court acknowledging that

> the wife's contributions and services are . . . substantial as a homemaker, spouse, mother of four sons, and manager of the households, both here and abroad. Her faithful devotion, her social companionship, intelligence, charm, friendships with the wives of his business contacts and linguistic fluency were all of importance to the husband in his worldwide social circles and in his career (p. 152).

The couple's story began in 1939, when they met at the Maisonette Russe Club in New York's St. Regis Hotel and fell in love. The wife was appearing as a singer and earning $500 per week. Although her future husband earned only $50 per week, he was the son of a prosperous British businessman. With his brother, the son set up a diamond merchandising firm in New York one year earlier.

The growth of this business, and its merger into Diamond Distributor's Inc., allowed the couple to enjoy a rather grand lifestyle of travel and entertainment in both New York and Paris. At the time of their divorce, the husband was president and CEO of Diamond Distributor's Inc., with a net worth exceeding $5 million, two-thirds of which was in company stock.

The Court determined that the stock was not marital property because its "increased value during marriage . . . [was] attributable to market factors" – especially the "diamond fever" of the late 1970s, when the couple were no longer together – and the husband's family associations (p. 147). This left as the principal assets to be divided real estate in Paris and New York.

The Court described both parties as "highly educated, intelligent and articulate" (p. 141). The wife was "a natural linguist, fluent in English, Spanish, French and thoroughly familiar with German, Portuguese and Russian" (p. 141), who had given up her promising career as an actress and singer in order to raise and care for the couple's four sons. The divorce was

Table 5.2. *Value of marital property in real divorce case*

Real estate	
90 percent of large apartment in Paris (family home)	$642,856
Paris studio	42,850
New York City coop (equity in)	103,079
1/3 of farm in Duchess County, New York	119,200
Liquid assets	
Cash and receivables	42,972
Securities	176,705
Corporate profit-sharing plan	120,940
Husband's life insurance (cash surrender value of)	24,500
Total	$1,273,102

a bitter end to their 41-year marriage, which was preceded by the husband's leaving his wife to live with another woman. The couple battled over (1) what constituted marital property, (2) how it was to be divided, and (3) the annual maintenance sum to be paid to the wife, with the case finally being decided in the appellate division of the Supreme Court in 1983.

We indicate in table 5.2 the value and type of assets that the Court deemed marital property. These assets do not include various articles of furniture, art, jewelry, and other personal property possessed by each party, and not claimed by the other, which the Court assessed as "about equal" in value and "already distributed" (p. 144).

The Court granted the husband all the real estate, except for the large apartment in Paris, and ordered that he pay the wife a series of three payments, over a span of three years, that would leave her with half the assets. While recognizing that "the wife seeks permanent exclusive use and occupancy of the Avenue Foch apartment [in Paris] as her marital home and/or a transfer of legal title to her" – to which she "has strong emotional attachment" (p. 145) – the Court ordered that it be sold after a period of three years from the date of its decision, saying it was too large, its upkeep too expensive (it generated over $43,000 in yearly expenses), and not needed for her four grown sons, who were by then self-supporting. This was a stunning setback for the wife.

By comparison, the husband was entitled to share in the sale of the Paris apartment as part of the marital distribution. The Court stipulated, however, that the husband contribute an annual maintenance sum of $65,000 to his wife.

This decision was appealed by the wife, but her appeal was denied. In his dissent in this case, 470 N.Y.S.2d 585 (A.D. 1 Dept. 1983), Justice J. Kassel wrote:

Table 5.3. *Point allocations of husband and wife in real divorce case*

Marital property	Husband	Wife
Paris apartment	35	55
Paris studio	6	1
New York City coop	8	1
Farm	8	1
Cash and receivables	5	6
Securities	18	17
Profit-sharing plan	15	15
Life insurance policy	5	4
Total	100	100

While it is true that the wife, now 67 years of age, is not infirm, it would be unfair in three years to compel her, then a 70 year old woman, to seek another home after having lived in the same apartment for more than 25 years of the 41 year marriage (p. 588).

Admitting that the trial court had reached "a very comprehensive and painstaking opinion" (p. 587), Justice Kassel nevertheless rejected the argument of his colleagues that "awarding former wife sole possession would materially upset delicate equitable distribution of assets meticulously worked out by the trial court" (p. 584). Instead, he said,

the wife should not be forced to vacate her home and submit to mandatory sale of the apartment. The entire ownership should be conveyed immediately to . . . [her] for her sole use, occupancy and disposition, in recognition of the various other residences owned by the parties which now remain with the [husband] (p. 587).

Recall that the trial court had awarded the three other pieces of real estate to the husband, on the grounds that the Paris studio met his need for a residence there for European business, and "the husband's equity in [the New York coop] cannot be recouped by forced sale" (p. 151 of appellate decision). The same seems to have been true of the farm, with its appraisal value based on a possible 1983 sale.

We surmise that the husband probably wanted to retain these three properties. By the same token, the lack of their mention in the wife's appeal leads us to conclude that she was interested solely in obtaining the family residence in Paris. As for the liquid assets, we assume that their value to both the husband and wife is roughly commensurate with their appraisal value.

On the basis of these assessments, the point allocations given in table 5.3, in which the larger of the two assignments is underscored, would seem to reflect the worth of the marital property to each party. By way of further explanation, the husband's point allocations indicate his greater interest in all the real estate except the Paris apartment, whereas the wife's allocations indicate the opposite preferences. Observe that both parties appraise the liquid assets either the same or within one point of each other.

The husband and wife win initially on their underscored items, which gives the husband only 45 points to the wife's 61 points. Giving the tied 15-point profit-sharing plan also to the wife (as specified in the description of AW), we see that the husband initially has 45 points and the wife has initially 76 points. Because the profit-sharing plan has the smallest ratio (15/15 = 1.0), it becomes the first item to be transferred in whole or in part. But even giving the entire 15 points to the husband only brings him to 60 whereas his wife still has 61. Hence, we must turn to the item with the second-lowest ratio, and this turns out to be the money and receivables which has a point ratio of 6/5 = 1.2. Thus, the wife must relinquish part of this so that equitability will be achieved.

The husband's and wife's total points will be equalized, with the husband getting the fraction α of the cash and receivables and the wife getting the fraction $1 - \alpha$ of this item, when:

$$60 + 5\alpha = 55 + 6(1 - \alpha).$$

(Notice that on the wife's side of the equation, we use 55 instead of 61, since this represents the number of points that she has, not counting the item that is involved in the transfer.) Solving for α gives $\alpha = 1/11$. Thus, the husband is entitled to 9.1 percent of the cash and receivables, or $3,906, and the wife $39,066, with each party receiving *in toto* 60.5 of his or her points (considerably fewer than the 69.0 points that each spouse received in the hypothetical case discussed in section 5.3).

Anticipating his wife's overwhelming desire for the Paris apartment – and that she would probably put the bulk of her points on it (as assumed in table 5.3) – the husband might consider (erroneously, it turns out) distributing his 100 points across everything else. If he won on everything else, which is likely, he would initially obtain 100 points to his wife's 55 points. But then he would have to give back several items, so he would actually come out worse than if he made the (honest) allocations shown in table 5.3. Hence, his undervaluing the Paris apartment, and overvaluing the other items, inevitably leads to his losing several of the latter, to his detriment.

On the other hand, the husband could benefit from almost matching his wife's 55 points on the Paris apartment (by putting, say, 54 points on it). Although he might lose about half the other items in doing so, all would have to be ceded back to him, plus some of the Paris apartment, in order to bring him up to his wife's point total.

But this deception strategy by the husband carries great risk unless he knows in advance exactly what his wife's point allocation will be. If he does not and puts, say, 56 points on the Paris apartment – thinking that his wife will put still more – both he and his wife could end up with property that only the other wants. In fact, such a bad call by the husband could bring his honest point total, as reflected in his point allocations in table 5.3, under 50 points if he ends up stuck with the Paris apartment (or most of it), which he values only at 35.

We conclude that attempts to manipulate AW, despite its vulnerability in theory (section 4.3), can badly hurt the players in practice – unless, of course, their intelligence about the other party's allocations is perfect or nearly perfect. But in the absence of a spy, such intelligence on the other party's allocations will in most cases be difficult to obtain and exploit. Hence, we think that AW will, in general, induce honest point allocations, even though, in the case of divorce, both parties presumably know each other intimately and might therefore be tempted to try to outguess each other.

A possible objection to the use of AW is that while it gives each party 60.5 of its points in the present case, it does not result in an equal division of the assets in terms of their monetary value. Thus, the wife gets $681,922 and the husband $591,180, so the wife would seem to receive a $90,742 bonus under AW. But this claim is mistaken, based on the points each party allocates to the items shown in table 5.3.

Recall that the monetary values are those estimated on the basis of appraisals made by the Court. But it is the points, which reflect the values of the players, that *should* count. In fact, to give back half of the wife's $90,742 bonus to the husband by giving him the cash and receivables and 1.0 percent of the Paris apartment, which are the lowest-ratio items, would create an imbalance in points: the wife would receive only 54.4 of her points, whereas the husband would receive 65.3 of his points – instead of their both receiving 60.5 points. Consequently, equitability would be destroyed to equalize the wrong thing – the Court's appraisal of the monetary value of the items, rather than the parties' own estimates of their worth.

Not only does the wife suffer by the forced sale of her Paris apartment within three years, but the husband also does not fare entirely well by the stipulations that the Court imposed on him. Although he was not forced to sell his vested interest in his company's profit-sharing plan, which was projected to more than double to $250,000 in five years, nor to surrender his life insurance policy for its cash value, he was directed by the Court to increase his life insurance coverage from $65,000 to $200,000 and make his wife the sole beneficiary. Also, he was ordered to continue to provide for her health insurance as well as maintenance.

This case is not unusual: one or both parties are often dissatisfied with the provisions of the settlement in divorce cases, with one-third to more than one-half of divorced individuals reporting that they were "seriously unhappy" with their settlements (Kressel, 1985, p. 12). When children are involved, the antagonistic negotiations of their parents may establish the terms of their future interaction, which, if rancorous, may leave the children with emotional or psychological problems.

5.5 AW versus Knaster's procedure

Recall that Knaster's procedure (section 3.2) assumes that players bid for items in an auction, and the highest bidder wins. The more successful players then make side payments to the less successful players – including those who won nothing – so that all players receive at least a proportional share in terms of items they won, money, or both. Thus, the initial losers are compensated for their losses not, as under AW, by giving them back goods but rather by monetary payments, based on the bids they made.

We showed in section 3.2 that Knaster's procedure, unlike AW, is not equitable – the players do not receive the same percentage increments above 50 percent. But this situation changes when the players all bid the same total amount, as we will next illustrate with the example used in section 4.3 to illustrate AW (and later PA).

Assume that the total valuations of Bob and Carol for the three goods, G_1, G_2, and G_3, are each \$100 (instead of the 100 points assumed under AW). As with AW, the underscored figures indicate who bids higher and who, therefore, wins each good:

	G_1	G_2	G_3	Total
Bob's valuation	$\underline{6}$	$\underline{67}$	27	100
Carol's valuation	5	34	$\underline{61}$	100

Thus, Bob receives G_1 and G_2, whose total valuation is \$73, and Carol receives G_3, whose valuation is \$61. Whereas goods were transferred (all of G_1 and 1 percent of G_2) from Bob to Carol to create equitability under AW, under Knaster's procedure each player's initial fair share is \$100/2 = \$50. Bob contributes \$73 – \$50 = \$23 to the surplus, and Carol \$61 – \$50 = \$11, making the surplus \$23 + \$11 = \$34. Splitting this surplus between them, each player's adjusted fair share is \$50 + \$17 = \$67, which is obtained by:

- giving Bob G_1 and G_2 and having him make a side payment of \$6 to Carol, which decreases his allocation from \$73 to an adjusted fair share of \$67;
- giving Carol G_3 and having her receive a side payment of \$6 from Bob, which increases her allocation from \$61 to an adjusted fair share of \$67.

Thus, Knaster's procedure results in not only an envy-free but also an equitable division, wherein both players receive 67 percent of their valuations, compared with 66 points under AW. The single percentage point under Knaster's procedure that is "lost" under AW is due to Carol's valuing less than

Bob the goods, G_1 and 1 percent of G_2, that are transferred to her. Notice, by the way, that none of this contradicts the efficiency of AW: because Knaster's procedure is allocating objects plus money, the contexts are not comparable.

In addition to being both envy-free and equitable when the two players have the same total valuations, Knaster's procedure is, as we saw in section 3.4, efficient. Recall that this follows from that fact that each player gets the items he or she values more than the other player; if compensation must be paid, it is done so at a cost less for a player than giving up the items he or she receives, which precludes exchanges of items or dollars between the two players that would make both better off.

In applying Knaster's procedure, the main practical difficulty lies in its assumption that players have money at hand that they can use to make the side payments necessary to achieve equitability. To be sure, if a player has a final deficit (as did Carol in our example) rather than a final excess (as did Bob), she does not have to spend her money, because she is the party being compensated rather than the one having to do the compensating. But, of course, there is no assurance beforehand that a player will be freed of the burden of having to make a positive side payment.

In the previous example, we artificially restricted the total bids of Bob and Carol to \$100 each (the total amounts do not really matter, only that they are the same). Let us now drop the restriction that Bob and Carol bid the same total amounts. Returning to the estate-division problem in Vermont we first described in section 1.2, suppose that not only do Bob and Carol value each of the ten items differently but also that they value the entire estate differently.

To be more concrete, assume that the bids Bob and Carol submit to the referee under Knaster's procedure are those shown in table 5.4.[8] We next offer a less formal explanation of the logic of Knaster's procedure than we did in section 3.2 for three players in order to highlight the intuition behind its calculations in this case.

Bob will think he is entitled to items worth half his total bid, or \$7,175/2 ≅ \$3,586, whereas Carol will think she is entitled to items worth half her total bid, or \$5,700/2 = \$2,850. Regardless of who receives what items, Knaster's procedure guarantees to each player at least this dollar amount in either items or money, with the money being sometimes added to what a player gets and sometimes subtracted from what a player gets.

The actual distribution of items and money is arrived at as follows: We begin by giving each item to the player who bid the most for it (with ties – such as that for the two mopeds – broken randomly), which are the bids underscored in

[8] These monetary values are consistent with the preferences that we supposed of the two players earlier. Different methods of valuing the assets of two players in estate division are compared in Weingartner and Gavish (1993).

Table 5.4. *Bids of Bob and Carol for Vermont estate*

Item	Bob's bid	Carol's bid
1 Boat	$1,000	$300
2 Motor	1,000	300
3 Piano	150	1,000
4 Computer	75	1,000
5 Rifle	300	200
6 Tools	150	300
7 Tractor	1,500	100
8 Truck	1,000	500
9 Moped	1,000	1,000
10 Moped	1,000	1,000
Total	$7,175	$5,700

table 5.4. Thus, Bob receives the boat, motor, rifle, tractor, truck, and (say) one moped, worth a total of $5,800 to him. He therefore thinks he received $5,800 – $3,586 = $2,214 more than what he is entitled to (his initial excess), so he turns this amount of cash over to the referee (temporarily). Carol, on the other hand, receives the piano, computer, tools, and one moped, worth a total of $3,300 to her. This is $3,300 – $2,850 = $450 more than she thinks she is entitled to (her initial excess), so she turns the $450 over to the referee, also. The referee now has $2,214 + $450 = $2,664 in cash, which is the surplus that he or she splits equally between the two players. Thus, Bob and Carol each receives $1,332 in cash back from the referee.

The final settlement finds Bob receiving items worth $5,800 to him and paying out a total of $2,214 – $1,332 = $882 in cash (the amount his initial excess is greater than his share of the surplus), whereas Carol receives items worth a total of $3,300 to her and receives $1,332 – $450 = $882 in cash (the amount her share of the surplus is greater than her initial excess). Hence, as in our earlier examples, the market clears. Notice that Bob thinks he received 68 percent of the estate (and Carol got 32 percent), whereas Carol thinks she received 73 percent of the estate (and Bob got 27 percent). Thus, neither player would trade what he or she received for what the other received, so the allocation is envy-free, as it must be when there are only two players, but it is not equitable because Carol received a higher percentage of her valuation than did Bob of his valuation.

The failure of equitability when the total bids of the players are different will make Carol presumably happier with the outcome. Indeed, if the bids are made known to each player, then Bob may resent Carol's better fortune – because she got more of what she wanted than he got of what he wanted – especially in light

Table 5.5. *Point allocations of Bob and Carol for Vermont estate*

Item	Bob	Carol
1 Boat	14	6
2 Motor	14	6
3 Piano	2	17
4 Computer	1	17
5 Rifle	4	4
6 Tools	2	6
7 Tractor	21	2
8 Truck	14	8
9 Moped	14	17
10 Moped	14	17
Total	100	100

of the fact that he bid the greater total amount and must make a side payment of $882 to her.

We next turn to AW, which never requires that the players make side payments since its adjustment mechanism occurs through the giveback of goods rather than the transfer of money. Compared with Knaster's procedure, AW is, we think, somewhat easier for the players to understand and apply. Among other reasons, making monetary bids under Knaster's procedure requires that the players compare the worth of an item not only with the other items but also with an absolute standard (i.e., money). Making point assignments under AW, by contrast, is less demanding in the sense that monetary bids can readily be translated into point assignments, but not vice versa.

To illustrate this translation, notice that Bob values the entire estate at $7,175. Since he also values the boat at $1,000, we know he thinks the boat represents $1{,}000/7{,}175 \cong 0.14$ of the estate. Thus, we have Bob assign 14 points to the boat. Continuing this conversion of dollars into points for each of the ten items yields the point assignments for Bob and Carol shown in table 5.5.[9]

In applying AW to the division of the Vermont estate, recall that AW begins by temporarily assigning each item to whoever puts the most points on it (underscored in table 5.5). Thus, Bob initially gets the boat, motor, tractor, and truck, giving him a total of 63 points. Carol receives the piano, computer, tools, and both mopeds, giving her a total of 74 points. Both value the rifle at 4 points, so it is awarded initially to Carol under the rules of AW, raising her total to 78.

[9] The individual point assignments may not always sum to 100 because of rounding, though they happen to do so here.

The rifle, however, is the first item transferred from Carol to Bob, which still leaves Carol with an advantage of 74 points to 63 + 4 = 67 points for Bob.

To prepare for the next transfer of items from Carol to Bob, we list the items Carol has in order of increasing quotients of Carol's points to Bob's:

one moped: 17/14 ≅ 1.2
one moped: 17/14 ≅ 1.2
piano: 17/2 = 8.5
computer: 17/1 = 17.0.

If we transfer one moped to Bob, his new total will be 67 + 14 = 81 points, and Carol's will be 74 – 17 = 57 points, indicating we have gone too far. Hence, we must calculate what fraction of the moped Carol must transfer to Bob in order for us to arrive at an equitable allocation, which will be efficient and envy-free as well.

Let α denote the fraction of the moped that will be retained by Carol. Then in order to equalize the point totals, thereby creating equitability, we must have

$$57 + 17\alpha = 67 + 14(1 - \alpha),$$

which yields $\alpha = 24/31 \cong 0.774$. Thus, Bob and Carol each receive 70.2 of his or her points – a 40 percent increment over half the estate – which is a greater proportion than in any of our previous examples.

Practically speaking, how do we transfer 77.4 percent of a moped from Carol to Bob? Perhaps they could reach an agreement whereby Carol uses it about nine months out of the year and Bob uses it the remaining three months. Perhaps not. A better solution might be that Bob announces what he considers to be a fair price for his 22.6 percent of the moped – say, $226, based on his bid of $1,000 under Knaster's procedure. Carol then chooses between buying Bob out at this price or selling her 77.4 percent share of the moped to Bob for 0.774/.226 ≅ 3.4 times the price he announced. Because, in this example, Carol also bid $1,000 for the moped and therefore values it the same as Bob, she will, in fact, be indifferent between buying Bob out at a price of $226 or selling out to him at a price of $774.

Like AW, Knaster's procedure is vulnerable to misrepresentation, as we showed in section 3.2. But for reasons already given, we believe that both procedures would be difficult to exploit in practice, especially if there are a relatively large number of items over which players must distribute their points.

Probably the most problematic feature of Knaster's procedure is its bankroll requirement; in many situations players will simply not have the money at hand that they need to make the side payments. Also, its rather complicated adjustment mechanism renders it less intelligible than the simple give-back-some-of-the-goods feature of AW that is used to create equitability, which

Knaster's procedure does not ensure unless the total amounts bid by the players are the same. On the other hand, Knaster's procedure is applicable to more than two players, which AW, at least in its two-person incarnation that preserves efficiency, envy-freeness, and equitability (see section 4.6), is not. Moreover, Knaster's procedure never requires the division of an item.

5.6 Conclusions

AW mimics some aspects of Knaster's procedure (section 3.2), except that Knaster's procedure requires that the players back up their bids with money, not just distribute points. Despite some attractive properties, Knaster's procedure seems rarely if ever to have been used. We suspect that the problem is not only one of having to possess money to use it but also that it is not nearly as simple as AW to understand.

True, the two procedures start off in the same way, with the larger bidder (in the two-person case) winning each item. But then the accounting under Knaster's procedure becomes quite complicated, whereas under AW the transfer of items back to the player who received fewer points initially is relatively straightforward.

In applying AW, probably the biggest problem is identifying a set of tolerably separable issues on which points are additive. As we illustrated in the case of the Panama canal treaty dispute, lumping nonseparable issues together leads to lower point totals for the players. Consequently, it is in the players' interest to try to identify as many issues as possible such that winning or losing on one does not greatly impinge on winning or losing on another.

We suggest that AW first be tried out in negotiations that involve easily specified issues or well-defined goods. Examples might include a dispute within a company over the division of job responsibilities, or the division of marital property in a divorce settlement, as we illustrated earlier. If the procedure works well in these settings, it might be used in more complex negotiations.

We focused on divorce settlements because of the sheer magnitude of the problem – half of all marriages end in divorce in the United States. AW, in our view, provides a straightforward settlement device that takes due account of the interests of both parties. Since the settlement is not the product of protracted negotiations or court battles, it is likely to lead to a more satisfying and durable outcome and foster more civil future relations between the parties, which is especially important if children are involved.

The fact that AW may circumvent litigation that drags on in court and drains husbands and wives of their resources may be a social good, but it will not delight lawyers if it robs them of legal fees. We believe, however, that lawyers can play a valuable role in AW's use by

- helping their clients make their point assignments in a way that reflects their honest estimates of worth, thereby enabling them to obtain better settlements;
- assisting them in predicting possible outcomes, perhaps by anticipating the allocations of the other party and running through various scenarios they might face, so as to reduce uncertainty.

Of course, it will remain for lawyers and courts to determine what constitutes marital property, to which AW can then be applied.

While honesty in making assessments will generally be the best policy, clients will still need to be assured of this. Also, they will need assistance in sorting out their feelings in order to make accurate assessments of value. The real divorce case (*Jolis v. Jolis*), in our opinion, illustrates how the litigation process can go wrong when judges insinuate their judgments into dispute resolution, thereby preventing the disputants from effectively expressing their own opinions, which we believe should be paramount in the process.

Just as a court's decision may be arbitrary, so the results of alternative dispute resolution (ADR) may be no better if ADR's relatively unstructured negotiations leave the parties without a procedure for reaching closure. AW seems to us to provide the discipline of formal proceedings that, at the same time, allow the parties – not just lawyers or judges – to speak for themselves. Furthermore, knowing they will be heard through the decisions about point allocations that they make should give them a greater sense of responsibility for their actions.

But divorce settlements are not the only domain in which the application of AW seems highly sensible. In the political arena, negotiations over arms control or border disputes often involve a plethora of issues that AW could help to resolve.[10] In the economic sphere, negotiations between labor and management over a new contract, or between two companies over a merger, are usually sufficiently complex that a point-allocation scheme could, we believe, prove very useful in finding a settlement that mirrors each side's most salient concerns.

Although we think it extremely unlikely that AW can be exploited in the absence of spies, for jittery negotiators the combined procedure might still be considered as an alternative. While it may make negotiators less nervous about starting with AW, however, it needs more careful analysis than we have given it here before it can be recommended as a serious alternative to the simpler AW.

[10] Hopmann (1991) illustrates the application of fair-division techniques to arms reductions by NATO and the former Warsaw Pact. Insofar as AW can be extended from two-person to n-person situations (section 4.6), it might well be used to assign cabinet posts to political parties in coalition governments so as to reflect the parties' entitlements based, for example, on their seat shares in parliament.

6 Envy-free procedures for $n = 3$ and $n = 4$

6.1 Introduction

Divide-and-choose, as we showed in section 1.2, assures each player of a piece of cake he or she perceives to be at least 1/2 the total (proportionality), no matter what the other player does. Because there are only two players, this means that each player can get what he or she considers to be a piece at least tied for largest. To provide this guarantee, however, the cutter must play "conservatively" by dividing the cake exactly in two, according to his or her valuation of it. That way, whatever piece the chooser selects, the cutter is assured of getting 1/2.

Although proportionality and envy-freeness are equivalent when there are only two players, this is not the case when there are more than two players. In extending envy-freeness to more than two players, we seek procedures wherein each player has a strategy that guarantees him or her a piece that is at least tied for largest, no matter what the other players do.

We showed in chapters 2 and 3 that none of the n-person proportional procedures is envy-free: while these procedures guarantee a player a piece of size at least $1/n$, one or more of the players might think that another player received a larger piece. This was not the case for the two-person point-allocation procedures we analyzed in chapter 4 and applied in chapter 5, which are not only proportional – and, therefore, envy-free – but also equitable.

Yet AW and PA, as well as the combined procedure, are strictly applicable only to the division of finitely many divisible goods. (However, if AW is used, only the good on which an equitability adjustment is made need be divisible.) This, of course, is a very different problem from that of dividing up a heterogeneous cake, which does not comprise homogeneous parts each of which can be separately allocated.

In this chapter we shall describe envy-free procedures for dividing up a heterogeneous cake in the cases of $n = 3$ and $n = 4$. Finding constructive envy-free procedures for five or more people – discussed in chapter 6 – has proved a formidable challenge for several decades.

One of the earliest results in the mathematical history of fair division is a proof that there exists an envy-free allocation of a cake among n persons if each person's preferences are given by what is technically known as a "countably additive measure" (Steinhaus, 1949). Steinhaus' proof, however, is only an *existence proof*: it provides "no clue as to how to accomplish such a wonderful partition" (Rebman, 1979, p. 33) and, moreover, relies on some advanced theorems (due to Lyapounov, 1940, and others) in mathematical analysis.

The first breakthroughs in solving the problem of obtaining envy-free allocations constructively – by specifying an algorithm for dividing up a cake, and then showing what strategies players can follow to guarantee envy-free portions for themselves – were made in the 1960s and 1970s for $n = 3$. In this chapter we shall present both these results and more recent results, including several new moving-knife algorithms for producing a three-person envy-free division, and one for producing a four-person envy-free division, that were discovered in the 1980s and the 1990s.

6.2 The Selfridge–Conway discrete procedure

The problem of finding a constructive scheme for producing an envy-free allocation among three players seems to have been first posed by Gamow and Stern (1958, pp. 117–19).[1] Such a procedure was found about 1960 by John L. Selfridge and, later but independently, by John H. Conway (Conway, May 17, 1993). Although never published by either, the procedure was quickly and widely disseminated by Richard K. Guy and others. Eventually it appeared in several treatments of the cake-division problem by different authors, including Gardner (1978), Woodall (1980), Stromquist (1980), and Austin (1982).

The Selfridge–Conway procedure is a discrete algorithm, involving at most five cuts, with up to three occurring in a first stage and two in a second stage. It requires that certain pieces at the first stage be combined with certain pieces at the second stage.

To describe the Selfridge–Conway procedure, we suppose there is a cake that three people (Bob, Carol, and Ted) wish to divide. They desire to achieve an envy-free allocation of the entire cake among themselves in a finite number of steps. To get to this point, we first show how they can obtain an envy-free allocation of *part* of the cake. We give both the procedure and the strategies that the players must employ to guarantee envy-freeness at each step:[2]

[1] This section is adapted from COMAP (1994, chapter 13) with permission. Nonconstructive results on the $n = 3$ case are given in Barbanel (1995c).

[2] We meld the rules of a procedure and the strategies of its players, which we distinguished in section 1.5, in order to make the description less choppy. It should be clear from the context which part of a statement is a rule (in step 1 below: "Bob cuts the cake into three pieces") and which part the strategy ("he considers to be the same size").

1 Bob cuts the cake into three pieces he considers to be the same size (this requires two cuts). He hands the three pieces to Carol.
2 Carol trims at most one of the three pieces so as to create at least a two-way tie for the largest (this requires a third cut). Setting the trimmings aside, she hands the three pieces (one of which may have been trimmed) to Ted. (The remaining steps now reverse the order of initial play.)
3 Ted now chooses – from among the three pieces – one that he considers to be at least tied for largest.
4 Carol chooses next – from the two remaining pieces – one that she considers to be at least tied for largest, with the proviso that if she trimmed a piece in step 2, and Ted did not choose this piece, then she must choose it now.
5 Bob receives the remaining piece.

Why is this distribution of part of the cake (the entire cake minus the trimmings) envy-free? Because Ted chooses first, he certainly envies no one. Since Carol created a two-way tie for largest, and at least one of these two pieces is still available after Ted selects his piece, she can choose one of these tied pieces and, therefore, will envy no one. Finally, Bob created a three-way tie for largest and, because of the proviso in step 4, the trimmed piece is not the one left over. Hence, Bob can choose an untrimmed piece and, therefore, will envy no one.

Before showing how this procedure can be extended to the entire cake, including the trimmings T, we illustrate the partial allocation just described. Suppose that all three players view the cake as having 18 units of value, with each unit of value represented by a small square. Suppose, however, that the players value various parts of the cake differently: Bob views the cake as being perfectly rectangular (2 x 9 in height and width), whereas Carol and Ted see it as being skewed to the right and left, respectively. We represent their different points of view pictorially as follows:

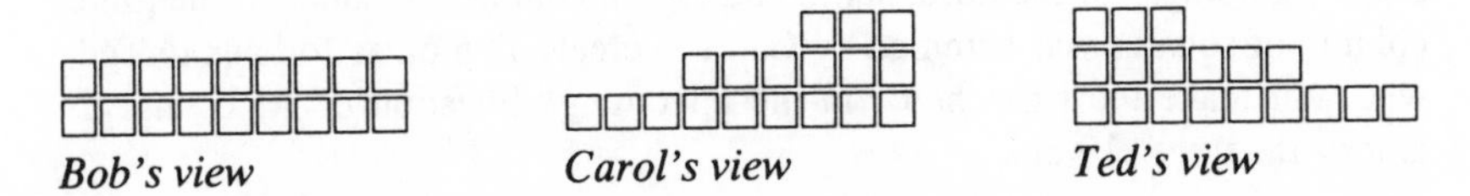

Bob's view *Carol's view* *Ted's view*

In step 1, Bob divides the cake into three pieces by making cuts at the right edges of the third and sixth columns from the left. This yields three pieces that he considers to be the same size (number of squares). From his point of view, this yields:

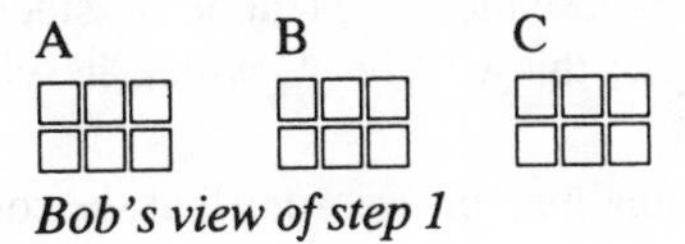

Bob's view of step 1

He now hands these pieces to Carol, who also views the vertical cuts as being made at the right edges of the third and sixth columns. She, however, sees the three pieces quite differently:

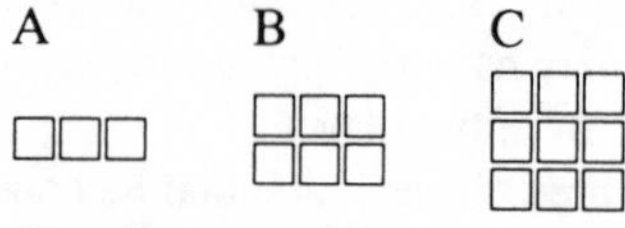

Carol's view of step 1

In step 2, Carol trims at most one of the three pieces so as to create at least a two-way tie for largest. In our example, this would mean trimming a piece from C to yield a piece C' that is the same size as B.

Although the most natural way for Carol to do this is to remove the top row of three squares from C, we will henceforth make all cuts in the vertical columns. (These cuts facilitate comparisons among players, as we will illustrate shortly.) Thus, step 2, from Carol's point of view, results in one column of three squares removed from C to give C'. This removed column we label with a T for "trimmings."

A B C' T

Carol trims T from C, yielding C', which is tied for largest with B

Carol now sets T aside and hands the three pieces (A, B, and C' in our illustration) to Ted.

To see how Ted views the results of the first two steps, we simply impose the vertical cuts that Bob and Carol made in steps 1 and 2 on Ted's view of the cake (which was the right-hand diagram in the first illustration). Their cuts at the right edges of the third, sixth, and eighth columns of squares (the ninth column of squares was trimmed by Carol to create T) look as follows to Ted, who will make the first choice in the upcoming allocation of A, B, and C' among the three players:

A B C' T

Ted's view of step 3

Among A, B, and C', Ted will now choose the piece that he considers to be at least tied for largest. In our example, this will be A, which he views as having 9 units of value.

Carol chooses next – from the two remaining pieces – one that she considers to be at least tied for largest, with the proviso that if she trimmed a piece in step 2, and Ted did not choose this piece, then she must choose it now. In our example, Carol thinks B and C' are tied for largest (see earlier diagram, in which Carol trims T), and both are available since Ted took A. Thus, the proviso dictates that Carol must choose C', which she views as having 6 units of value.

Finally, Bob receives the remaining piece, which is B in our example and which gives him 6 units of value (see first diagram). (Notice that, without the proviso, Bob could have been stuck with the trimmed piece, C', which he thinks is only 4 units of value since its third column was trimmed off by Carol.) In sum, Bob gets 6 of his units (1/3 of the cake), Carol gets 6 of her units (1/3 of the cake), and Ted gets 9 of his units (1/2 of the cake). Thus, even before touching the trimmings T, each of the players, at least in our example, has received a proportional share (i.e., 1/3) of the entire cake.

Having obtained an envy-free allocation of the cake – minus the trimmings T that Carol cut off – the players must concern themselves with what to do with T. It turns out that T can be allocated in such a way that the resulting allocation of the *entire* cake is envy-free. An explanation of how this can be done will complete our description of the Selfridge–Conway procedure.

The key observation in allocating T among the three players is that Bob will not envy the player (assume it is Ted) who receives the trimmed piece, even if Ted were to be given all of T (although that is not actually what we will do). The point is that Bob created a three-way tie among the pieces and received an untrimmed piece, while the trimmed piece – even with T added onto it – would yield Ted only a piece that Bob considers to be exactly the same size as the one he (Bob) received. We describe this situation by saying that Bob has an *irrevocable advantage* over Ted.

To illustrate how the trimmings are now allocated, let us continue to assume that it is Ted who received the trimmed piece (it could as well be Carol). Thus we are assured that Bob has an irrevocable advantage over Ted.

We now let Carol cut T into three pieces she considers to be the same size (these are the fourth and fifth cuts of the Selfridge–Conway procedure).[3] Let the players choose which of the three pieces they want in the following order: Ted, Bob, Carol.

To see that this yields an envy-free allocation of the entire cake, notice that Ted envies no one, because he chooses first in the division of T and can take the piece he considers largest. Bob, choosing next, does not envy Carol, because he is choosing ahead of her in the allocation of the trimmings; and he does not envy Ted, because, as we said earlier, Bob has an irrevocable

[3] In general, it is whichever player did *not* get the trimmed piece in the earlier allocation who now plays the role of divider of T.

advantage over Ted and, hence, would not envy him regardless of how the trimmings were allocated. Finally, Carol envies no one, because she made all three pieces of T the same size.

To recapitulate, we began with an envy-free allocation of all the cake except T. We followed this by an allocation of T, which yields an envy-free allocation of the entire cake.

In our earlier example with the small squares, Carol will divide T into thirds, each third being worth 1 unit to her. This equal three-way division will translate into three 2/3-units for Bob, and three 1/3-units for Ted, with each player receiving one of his or her fractional units. *In toto*, the players receive the following allocations:

$$\text{Bob: } 6 + 2/3 = 6^2/_3; \text{ Carol: } 6 + 1 = 7; \text{ Ted: } 9 + 1/3 = 9^1/_3.$$

If either Carol or Ted rather than Bob were the first divider, then the allocation of the cake on the first round would be different from that which we gave earlier (each player would receive some fractional units on this round). The trimmings, also, would be different, yielding a different envy-free allocation of the entire cake. The fact that there may be different envy-free allocations, depending on which player is the first divider, suggests that not all allocations will be equally desired by each player.

We will take up the question of the overall desirability of different allocations in section 7.6. For now it is worth pointing out that an obvious envy-free allocation in our example is to give to players the columns they most value: Carol the right-hand three columns (9 units), Ted the left-hand three columns (9 units), leaving the middle three columns for Bob (6 units). Interestingly enough, two of the three players (Bob and Ted) would prefer the Selfridge–Conway allocation – in which Bob is the first divider – to this allocation, so this envy-free allocation is perhaps not as "obvious" a solution as it would first appear.

6.3 The Stromquist moving-knife procedure

Probably the best-known but most complicated moving-knife procedure is the envy-free procedure for three people due to Walter Stromquist (1980).[4] This procedure begins with a referee holding a knife at the left edge of the cake, and each of the three players holding a knife parallel to the referee's at a point that that player thinks exactly halves the cake.

The referee moves his or her knife slowly across the cake, as was the case with all the previous moving-knife procedures. Each player keeps his or her

[4] This section and the next two sections are adapted from Brams, Taylor, and Zwicker (1995b) with permission.

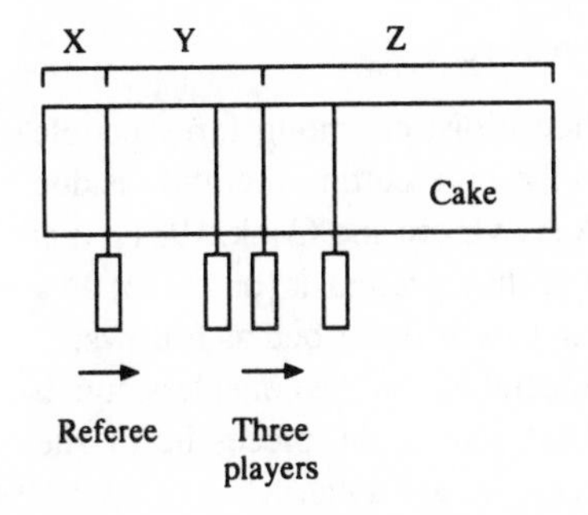

Figure 6.1 Stromquist moving-knife procedure

knife to the right of the referee's knife and parallel to it, always positioning it so that it exactly halves – in the player's estimation – the piece to the right of the referee's knife.

At any time, a player can call "cut"; he or she then receives the piece to the left of the referee's knife (X in figure 6.1). A cut is then also made by whichever of the three players' knives is in the middle (yielding Y and Z). Of the other two players, the one whose knife is closest to the referee's gets Y, and the other gets Z.

The strategy is for a player to call cut only if he or she thinks the left-hand piece X is at least as large as both the middle piece Y and the right-hand piece Z.[5] Hence, the player calling cut (assume it is Bob) will never envy the other two (Carol and Ted). Since neither Carol nor Ted called cut, they must each think the largest piece is either Y or Z.

To see that neither of them experiences any envy, notice that whoever received Y (assume it is Carol) held either the second or third knife from the left (counting the referee's) in figure 6.1. Hence, Carol thinks that Y is either strictly larger than Z (if Carol's knife is the second knife from the left), or tied with Z for largest (if Carol's knife is the third knife from the left). Similarly, Ted, who received Z, held either the third or fourth knife from the left. Hence, he thinks that Z is either tied with Y for largest (if Ted's knife is third from the left) or strictly larger than Y (if Ted's knife is the fourth knife from the left).

Stromquist (1980, p. 641) notes that his procedure, like the Selfridge–Conway procedure, does not generalize to $n > 3$. Neither, it appears, does the next procedure.

[5] Misinterpretations of this strategy have caused some confusion in the literature; see Stromquist (1981), Austin and Stromquist (1983), and Olivastro (1992b).

6.4 The Levmore–Cook moving-knife procedure

Another procedure for producing an envy-free division among three people that seems to have been largely overlooked in the cake-cutting literature is due to Saul X. Levmore and Elizabeth Early Cook (Levmore and Cook, 1981). It is essentially a moving-knife algorithm, although they present it (on p. 52) as a process with "infinitely small shavings." It can can be described as follows:

Bob divides the cake into three pieces, P, Q, and R, that he considers equal. Each of the other two players (Carol and Ted) selects the pieces he or she considers largest. If they choose different pieces, we are done.

Otherwise, we can assume that they both choose P. Now Bob starts a vertical knife moving rightward, as in the Dubins-Spanier moving-knife procedure (section 2.4), but at the same time he places a second knife, perpendicular to the first, over the portion of the cake over which the first knife has already swept (see figure 6.2).

Notice that if cuts were to be made from such a positioning of the two knives, the piece of cake labeled P would be cut into three pieces, exactly two of which would involve both knives (i.e., the pieces to the left of the vertical knife). Let S denote one of these two pieces and let T denote the other. Assume Bob moves the horizontal knife up and down in such a manner that he thinks Q together with S is the same size as R together with T.

When the process begins, both S and T are empty, so Bob thinks Q and R are the same size. Carol and Ted, on the other hand, both think that P, with S and T removed (both are initially empty), is larger than both Q together with S and R together with T. This gives us four inequalities at the start of play:

- Carol's size of P with S and T removed > Carol's size of Q plus S.
- Ted's size of P with S and T removed > Ted's size of Q plus S.
- Carol's size of P with S and T removed > Carol's size of R plus T.
- Ted's size of P with S and T removed > Ted's size of R plus T.

Now let either Carol or Ted call stop when one of these four inequalities first becomes an equality. Without loss of generality, assume it is Carol, and this is because she now thinks that P with S and T removed is equal to R together with T. At this point we can obtain an envy-free division by giving:

- Carol R together with T;
- Ted P with S and T removed; and
- Bob Q together with S.

Levmore and Cook (1981, p. 52) suggest an extension to four players:

> The four daughters are A, B, C, and D. Have A divide the land into four plots such that she would be happy with any of them. Call the plots 1, 2, 3, and 4. The other three daughters indicate their preferences. Needless to say we are stuck with somewhat similar tastes for the purpose of this problem. Perhaps B and C settle into plot 1 and D chooses 2. Now A must shave from 1 into 2, 3, and 4. (We might have had the shaving

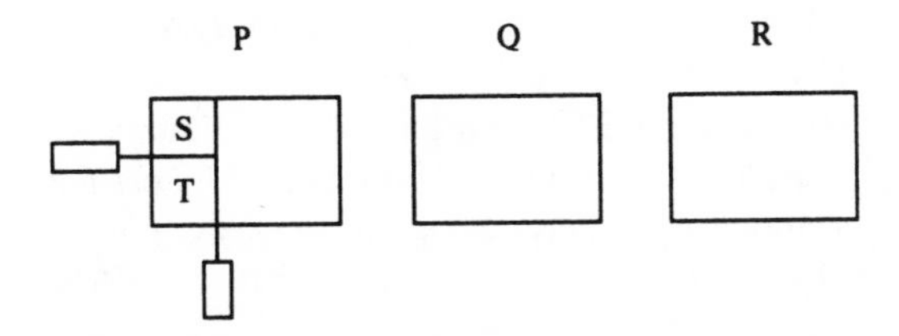

Figure 6.2 Levmore–Cook moving-knife procedure

done into 3 and 4, figuring that we can leave well enough alone in plot 2. However, as we increase 3 and 4 we can count on D's increasing jealousy and her reentry, so we take this into account at the outset.) A continues the shaving until B or C moves into a new plot. Whenever two daughters are in the same plot or three in the same plot), as when B moves into 2 in competition with D, A turns her attention to that plot and shaves from it into all the others (from 2 into 1, 3, and 4).

We next demonstrate why this procedure will not always work. By construction, A values the plots the same – at 1/4 each. Assume that B thinks plots 1 and 2 are worth 5/8 and 3/8, respectively. Suppose that C thinks that plot 1 has all the value (1), and D thinks that plot 2 has all the value (1). In summary, the players value plots 1, 2, 3, and 4 as follows:

	Plots			
	1	2	3	4
A	1/4	1/4	1/4	1/4
B	5/8	3/8	0	0
C	1	0	0	0
D	0	1	0	0

Suppose that on the first trimming (shaving) by A, she chooses subsets X, Y, and Z of plot 1 that A, C, and D think are worthless (0 value) and transfers these to plots 2, 3, and 4, respectively. Assume B thinks X is worth 1/4 and Y and Z have 0 value. The result is that B will move to plot 2, because her new valuation is as follows (the other players' valuations stay the same):

	1	2	3	4
B	3/8	5/8	0	0

Now this situation is exactly the same as when we started, except that the roles of plots 1 and 2 have been reversed for B. Thus, the Levmore–Cook procedure, at least as described above, could have B moving back and forth between plot 1 and plot 2 with no convergence – except for the transfer of sets, worthless in value to all other players, from plots 1 and 2 to plots 3 and 4. We know of no way to make a moving-knife version of this procedure work, as we were able to do for the $n = 3$ case.

Saul X. Levmore is a law professor and Elizabeth Early Cook a journalist. Their idea for the $n = 3$ case seems to us an imaginative one, even if it does not generalize. Under our interpretation, it has only two simultaneously moving knives, which is unquestionably simpler than Stromquist's (1980) procedure with four simultaneously moving knives.

6.5 The Webb moving-knife procedure

It turns out that by combining the basic idea in the Dubins–Spanier moving-knife procedure (section 2.4) with Austin's moving-knife procedure (section 1.5), one can obtain a fairly simple moving-knife procedure that guarantees an envy-free allocation among three people. This procedure was first discovered by William A. Webb (n.d.), although he was unaware of Austin's work and thus recreated the part of it he needed. The version we give next uses Austin's original procedure:

A knife is slowly moved across the cake, as in the Dubins–Spanier procedure, until some person – assume it is Bob – calls cut (because he is the first to think the piece so determined is of size 1/3). Call the piece resulting from this cut A_1, and notice that Carol and Ted both think A_1 is of size less than 1/3 (and thus they think the rest of the cake is of size greater than 2/3).

We now have Bob and either one of the other two players – assume for definiteness it is Carol – apply Austin's procedure to the rest of the cake, resulting in a partition of it into two sets, A_2 and A_3, that Bob and Carol agree is a 50–50 division of the rest of the cake. Notice that:

1 Bob thinks that all three pieces are of size exactly 1/3.
2 Carol thinks A_2 and A_3 are tied for largest (since she views each as being exactly 1/2 of a piece that is more than 2/3 of the whole cake).

An envy-free allocation is now easily obtained by having the players choose among the three pieces in the following order: Ted, Carol, Bob. Note that Ted envies no one because he is choosing first; Carol envies no one because she had two pieces tied for largest (at most one was chosen by Ted); and Bob envies no one because he thinks all three pieces are the same size.

There are obvious generalizations of Webb's procedure to n players, but their success depends on corresponding generalizations of Austin's procedure that are not – at the time of this writing – known to exist. For example:

> If there exists a moving-knife procedure P that will divide a cake into $n - 1$ pieces so that each of $n - 1$ players thinks all the pieces are of size $1/(n - 1)$, then there exists a moving-knife procedure for producing an envy-free division of the cake among n players.

To see how the envy-free procedure would work for n players, we begin, as in the Dubins–Spanier procedure, by obtaining a piece of cake that, say, Bob thinks is of size exactly $1/n$ and everyone else thinks is of size at most $1/n$. We now have Bob, together with any $n - 2$ of the other players, divide up the rest of the cake into $n - 1$ pieces using P so that each player thinks that the $n - 1$ pieces are all the same size. The one player not involved in the application of P then gets to choose first, whereas Bob is forced to choose last. The order in which the other players choose is immaterial. Envy-freeness follows in the same way that it did for Webb's procedure.

If $n = 4$, this argument shows that a four-person envy-free procedure can be obtained from an extension of Austin's procedure that would allow three players to divide a cake into three pieces that each thinks is of size exactly 1/3. This approach, however, is unnecessary: we give in section 6.7 a moving-knife solution to the four-person envy-free problem recently found by Zwicker and the authors (Brams, Taylor, and Zwicker, 1995a).

On the other hand, no one has yet found a moving-knife procedure that will yield an envy-free allocation among five players. The argument we gave based on Webb's procedure shows that such a procedure would exist if one had an extension of Austin's procedure that would allow four players to divide a cake into four pieces that each thinks is of size exactly 1/4. Of course, if the four players could always get a division of the cake into two pieces that each agreed was 50–50, then they could apply this procedure to each of the two pieces to obtain the desired extension of Austin's procedure. Thus, the argument behind Webb's procedure shows the following:

If there exists a moving-knife procedure P that will divide a cake into two pieces so that each of four players thinks it is a 50–50 division, then there exists a moving-knife procedure for producing an envy-free division of the cake among five players.

At the expense of a more complicated argument – which is, however, not without its charms – this result is improved upon in Brams, Taylor, and Zwicker (1995a) to show that all one needs is an extension of Austin's procedure to three players (not four) in order to obtain a moving-knife solution to the five-person envy-free cake-division problem. More on this, and a formalization in terms of "oracles," can be found in Brams, Taylor, and Zwicker (1995a).

6.6 Two other moving-knife procedures for $n = 3$

A conceptually simple envy-free moving-knife procedure for three players can be achieved by picturing a round cake (or pie, as in Gale, 1993) and using radial knives. The following procedure, as far as we know, originated in Brams, Taylor, and Zwicker (1995a).

Start by having Bob hold three knives over the round cake as if they were hands of a clock in such a way that he considers the three wedge-shaped pieces to be all of size 1/3. Now have Bob start moving all three knives in a clockwise fashion so that each piece remains of size exactly 1/3 in his valuation, subject to the requirement (superfluous if the aforementioned strategies are followed) that the moment any knife reaches the initial position of some other knife, all three knives line up with the initial positions.

The claim is that, at some point, Carol must think at least two of the wedges are tied for largest. That is, if Carol thinks a single wedge (call it A) is largest at the instant when the knives start moving, then A is eventually transformed, as in Austin's procedure, to the wedge immediately clockwise. At some point prior to this, the piece between the two knives that originally defined A loses its position as largest to another piece (either that immediately clockwise or immediately counterclockwise), at which point Carol calls "stop." At this instant, we have the desired two-way tie for largest in Carol's eyes.

The envy-free allocation is now obtained by having the players choose in the following order: Ted, Carol, Bob. This procedure can be recast as one in which three knives move in parallel across a rectangular cake, with the understanding that as a knife slides off the right edge, it immediately jumps back onto the left edge.

Another envy-free moving-knife procedure for three people is an immediate consequence of a further generalization of Austin's procedure for obtaining a piece of cake that each of two players thinks is of size exactly 1/3. This generalization, described in section 1.5, yields a partition of the cake into k pieces so that each of two players thinks all k pieces are of size $1/k$.

To obtain the envy-free procedure from this generalization of Austin's procedure, one simply has, say, Carol and Ted use Austin's procedure to obtain a partition of the cake into three pieces they both think are all of size 1/3. The players then choose the piece they want in the following order: Bob, Carol, Ted (the order of Carol and Ted can be reversed). Bob experiences no envy because he is choosing first; and neither Carol nor Ted will experience envy because each thinks that all three pieces are the same size.

The real benefit of the last three-person envy-free procedure is that the idea behind its first step – having two players simultaneously create an equal division of the cake into as many pieces as there are players – is the key to solving the corresponding four-person problem. We turn to this problem and its solution next.

6.7 A four-person moving-knife procedure

The four-person moving-knife procedure that we describe next – due to William S. Zwicker and the present authors (see Brams, Taylor, and Zwicker,

1995a) – makes use of Austin's procedure, but this time to have Bob and Carol divide a cake into four pieces that they both agree are all of size 1/4. Ted now trims at most one of the four pieces to create at least a two-way tie for largest in his estimation. The trimmings, which will be allocated shortly, are set aside.

We can achieve an envy-free allocation of all the cake except the trimmings by having the players choose among the four pieces (one of which may have been trimmed) in the following order: Alice, Ted, Carol, Bob, subject to the proviso that Ted must take the trimmed piece if there is one and Alice did not take it. Notice that neither Carol nor Bob will envy anyone because they receive untrimmed pieces and think all four of the original pieces are the same size.

For the allocation of the trimmings, we make use of the same idea as used in the Selfridge–Conway discrete procedure (section 6.2), but now we invoke Austin's procedure. The key is that Bob (and Carol, although we will not need this fact) have an irrevocable advantage over whichever player took the trimmed piece in the partial allocation. That is, Bob will not envy the player with the trimmed piece, regardless of how the trimmings get allocated.

As for Ted and Alice, let the one who took the trimmed piece be renamed "the noncutter," the other "the cutter." Thus, Bob has an irrevocable advantage over the noncutter. (These names will make sense in a moment.)

For the allocation of the trimmings, we have the cutter and Carol invoke Austin's procedure to divide the trimmings into four pieces that they agree are all the same size. The players then choose among these four pieces, with the noncutter choosing first, Bob second, and then the cutter and Carol in either order.

The resulting allocation of the whole cake is envy-free. The noncutter envies no one, because he or she is choosing first in the allocation of the trimmings. Bob envies neither the cutter nor Carol, because he is choosing ahead of them. Moreover, Bob does not envy the noncutter – even though the noncutter is choosing ahead of him – because he (Bob) had an irrevocable advantage over the noncutter. Finally, neither the cutter nor Carol will envy anyone, because the four pieces into which they cut the trimmings are all of equal size, so the players who choose before them gain no advantage.

This four-person moving-knife procedure may be viewed as a generalization of the Selfridge–Conway three-person discrete procedure (section 6.2). By creating a four-way tie for Bob and Carol initially under the former procedure, rather than a three-way tie for Bob alone under the latter procedure, the four-person procedure enables one to add an extra player. The second stages of the two procedures for creating an irrevocable advantage are also analogous, with the trimmings under the four-person procedure distributed among four players rather than the three persons under the Selfridge–Conway procedure.

Not only these two procedures, but all the other envy-free procedures discussed in this chapter, are *bounded.* That is, there is a finite upper bound on the number of cuts that are needed to construct, via the algorithms, an envy-free division. We saw that at most five cuts are required in the case of the Selfridge–Conway algorithm (section 6.2), and at most eleven cuts are needed in the case of a slight modification of the four-person procedure (Brams, Taylor, and Zwicker, 1995a).

The four-person envy-free procedure does not seem to generalize in any obvious way to the case of five or more persons. However, as pointed out in section 6.5, it is possible to show that if there were an algorithm that enabled three people to agree on a 50–50 division of a cake – instead of two people, as allowed under Austin's procedure – then it would be possible to extend the four-person bounded moving-knife procedure to a five-person bounded moving-knife procedure (Brams, Taylor, and Zwicker, 1995a).

In chapter 7 we will discuss a discrete algorithm which produces an envy-free allocation for any number of players. Interestingly enough, it is based on some of the same ideas (trimming and irrevocable advantages) found in both the Selfridge–Conway three-person envy-free procedure and the four-person envy-free procedure.

7 Envy-free procedures for arbitrary *n*

7.1 Introduction

In this chapter we present the solution to the problem first raised by Gamow and Stern (1958) of finding a finite algorithm for providing an envy-free division among any number of players. Before presenting this solution, however, we will describe two algorithms in section 7.2 – one continuous, involving a moving knife, and the other discrete – that provide allocations that are approximately envy-free, with the degree of error within any preset tolerance level.

We describe in section 7.3 an exact but "infinite" solution to the n-person envy-freeness problem, using what we call the "trimming procedure." Because this procedure requires an infinite number of stages to accomplish, however, it is not, technically, an algorithm.

Fortunately, the trimming procedure can be rendered finite, but only by complicating it considerably. Without giving full details, we will describe enough of the finite procedure in section 7.4 to give the reader a good idea of how it works, and why it is unbounded – the number of stages it may require cannot be specified without knowing the preferences of the players. We will also illustrate this procedure with an example involving four players.

In section 7.5 we apply the trimming procedure to an inheritance example that involves indivisible as well as divisible goods, showing that it may be necessary to sell some of the indivisible goods to complete the procedure. We indicate in section 7.6 how it and other envy-free procedures can be applied to chore division, and how entitlements can be incorporated into some of the envy-free procedures.

Like the n-person proportional procedures described in this and the previous chapter, the envy-free procedures do not generally yield efficient allocations. Under special conditions we describe in section 7.6, however, we will demonstrate that a three-way division of a cake is efficient in a weak sense.

7.2 Two approximate procedures

We begin by altering the Dubins–Spanier moving-knife procedure (section 2.4) to allow a player to reenter the process in some way. One possibility that suggests itself is to allow a player to call "cut" again and again, even though he or she already has done so at least once and thus received a piece of cake. That player would then be required to take the new piece, determined by this most recent cut, and return his or her previous piece to the cake.

What this yields is the following: Given n players and some small positive number ε, there is a moving-knife procedure that will guarantee each player a piece of cake that he or she thinks is smaller, by at most ε, than the largest piece. The procedure is simply the one we just described, with the appropriate strategy being for each player initially to call cut whenever he or she thinks the piece this will yield is of size $1/n$, and thereafter to call cut whenever he or she thinks the new piece is larger by ε than the one he or she presently holds. Ties are broken at random.

For the reader who is concerned that the rules of this procedure (as opposed to the strategy described) would allow a player to call cut infinitely many times, we remark that it is easy to see that the strategy described is not affected by an additional rule that asserts that each player can call cut at most $1/\varepsilon$ times. Thus, if $\varepsilon = 1/100$ of the cake, a player need never call cut more than 100 times to ensure that his or her piece is "almost" the largest – that is, smaller than the largest piece by at most 1/100 of the entire cake. Hence, as long as ε is positive, no matter how small, the process will eventually terminate after a finite number of cuts.

The approximation, as determined by ε, may be to as high a degree of accuracy as one desires. Indeed, when ε becomes sufficiently small, a piece that is ε bigger than another would – in a real-world application – become indistinguishable from it.

Of course, different players may have different ε's, but for simplicity we will assume that all the players have the same ε. Let us call this amount the *minimal distinguishable difference* between pieces: any piece larger than another piece by ε (or more) will be seen by a player as preferable, giving him or her reason to call cut again to relinquish the smaller piece and acquire the larger piece. But a piece that is bigger than another, but by less than ε, will not be perceived as bigger, making the two pieces indistinguishable.

The perceptions of the players become the reality in real-world applications. Thus, procedures that yield allocations which are approximately envy-free may be as valuable as the exact solutions we shall present later.

It turns out that there is also a discrete procedure for obtaining the kind of almost envy-free allocation just described. Moreover, it is of additional interest because it also provides an approximation to the two-person version of

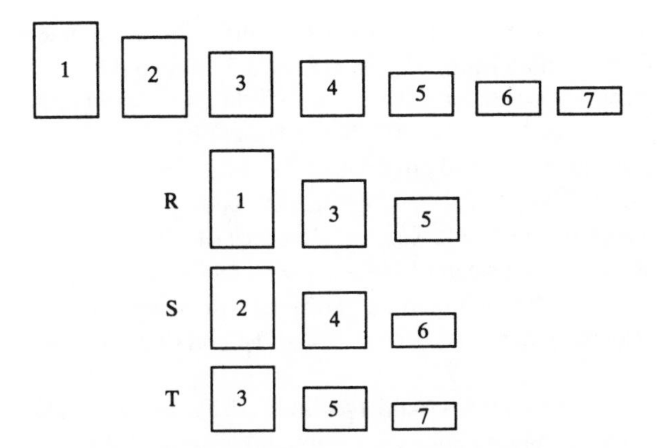

Figure 7.1 Envy-free approximate discrete procedure

Austin's moving-knife procedure, described in section 1.5. (Recall that this procedure yields a partition of the cake into k pieces in such a way that both players agree that each of the k pieces is of size exactly $1/k$.) A very different way of achieving an approximation to Neyman's (1946) result was known to Jack Robertson and William Webb.

The basic idea of our approximate discrete procedure is most easily conveyed through a simple example (Taylor, 1993). Suppose that two people wish to divide a cake into two pieces so that each thinks it is essentially a 50–50 split. By "essentially" we will, for now, mean "within 15 percent." Here is how that can be accomplished:

We begin by having Bob divide the cake into seven pieces he considers to be the same size. Carol, in general, will see them as unequal; assume she arranges them from largest to smallest and numbers them 1 (largest) to 7 (smallest). Now look at three separate chunks of cake, labeled R, S and T, which combine different pieces (see figure 7.1). We assume that

- R is made up of pieces numbered 1, 3, and 5;
- S is made up of pieces numbered 2, 4, and 6;
- T is made up of pieces numbered 3, 5, and 7.

Notice first that R and S are disjoint, so Carol cannot think both are more than 1/2 of the cake. In particular, the smaller of these two chunks cannot be more than half the cake. But the smaller chunk must be S, because 2 is smaller than 1, 4 is smaller than 3, and 6 is smaller than 5, at least in Carol's estimation. Thus, we have established that Carol thinks S is at most half the cake.

Carol also thinks T is no larger than S (since 2 is larger than 3, 4 is larger than 5, and 6 is larger than 7). Hence, she thinks T is also at most half the cake.

Notice that S and T together comprise all the cake except for the one piece numbered 1. Carol can thus take this single piece and distribute it between S and T so that each is brought up to size exactly 1/2 in her estimation. Of course, from Bob's point of view, S and T were of size exactly 3/7, and so the most uneven the split can become – whatever Carol does with the piece numbered 1 – is to make one chunk 3/7 and the other 4/7. Since $1/7 \cong 0.143$ is less than 15 percent of the cake, we have achieved our goal.

Sticking with the case of two players for a moment, it is easy to see how we can generalize what we have done to handle tolerances of different size. Thus, if we start with $2n + 1$ sets instead of 7, then a construction completely analogous to that we just described will yield a partition of the cake into two pieces that one of the players thinks is an even split and the other thinks is off by at most $1/n$.

Suppose, for example, that we want to obtain a partition of the cake into, say, nine pieces that each of two players thinks is of roughly size 1/9. We can proceed as follows: Using what we already know, we can divide the cake into two pieces that both players think is essentially an even split. Then we can do the same to each of these two pieces, arriving at a partition of the cake into four pieces that both players think are essentially the same size. Continuing in this manner, we will eventually get to a point where we have 2^n slivers that each player thinks is of size about $1/2^n$.

Now when n is sufficiently large, each player will think that each of the individual slivers is smaller than, say, half the tolerance that was set. Assume, for example, that this happens when $n = 7$, and we thus have 128 slivers. We can then obtain the desired nine pieces by grouping these 128 slivers into nine blocks, each of which consists of either 14 or 15 of the 128 slivers. Any deviation from this allocation's being an equal division in the eyes of either of the two players will be caused by:

- the approximations introduced in getting to the 128 slivers (which we can make as small as necessary (e.g., half the tolerance that was set); or
- the fact that $14 < 128/9 < 15$ (and we made the slivers sufficiently small so that this would be within half the given tolerances).

What we have done so far gives us a partition of the cake into as many pieces as we want such that each of two players will think all the pieces are essentially the same size. How can we now handle the case of three people? The answer lies in going back and redoing everything we just did for two people (Bob and Carol), but with "Bob" now replaced by "Bob and Ted."

The point is that all Bob did earlier is to divide the cake into many pieces that he considered to be of the same size. But instead of Bob's being the sole divider, we can start with a division of the cake into many pieces that both Bob and Ted agree are essentially the same size. Following through on this yields a partition of the cake into as many pieces as we want such that each of three players will think all the pieces are essentially the same size.

How can we now handle the case of four people? Again, we go back and redo everything we just did for three people, but with "Bob and Ted" replaced by "Bob and Ted and Carol." And so on.

7.3 An infinite procedure

A naive attempt to generalize to $n = 4$ what the Selfridge–Conway procedure (section 6.2) does for $n = 3$ would begin as follows: Bob would cut the cake into four pieces he considers to be the same size.[1] Then Carol and Ted would trim some pieces (but how many?) to create ties for the largest. Finally, the players would choose from among the pieces, some of which would have been trimmed, in reverse order: Alice, Ted, Carol, Bob.

Alas, this approach fails because Bob could be left in a position of envy. In order to understand how such a failure can occur, consider how many pieces Ted would have to trim in order to create a sufficient supply of pieces, tied for largest, so that he is guaranteed to have one available when it is his turn to choose.

It would appear that Ted might have to trim one piece to create a two-way tie for largest. That way, whichever piece Alice subsequently chose, there would be at least one other largest piece that Ted could choose next.

Similarly, it would appear that Carol would have to trim two pieces to create a three-way tie for largest. That way, whichever two pieces Ted and Alice subsequently chose, there would be at least one other largest piece that Carol could choose next.

Label the four equal-size pieces that Bob cuts initially A, B, C, and D. A problem that arises with this approach occurs when

- Ted regards B as the largest piece and trims it to B' in order to tie with his next-largest piece, A; and
- Carol regards C and D as her two largest pieces and trims them to C' and D' in order to tie with her third-largest piece, also A.

Now if Alice chooses A, it does not matter what pieces Carol and Ted choose (B', C', or D'): there will not be an untrimmed piece left for Bob, so he will have to take a trimmed piece and, consequently, will envy Alice for having taken the only untrimmed piece, A.

But all is not lost, because there are modifications in the first stage of the Selfridge–Conway procedure that we can make for arbitrary n to obtain a partial allocation of the cake (the rest of the allocation will be completed in later stages). We describe one such modification for $n = 4$ and another later in the section. The new idea we introduce is to have Bob cut the cake into *more* pieces than there are players. The procedure, and the strategies that the players can use at each step to guarantee an envy-free division of part of the cake, are

[1] This section is adapted from COMAP (1994, chapter 13) with permission.

as follows:

1 Bob cuts the cake into *five* pieces that he considers to be the same size. He hands the five pieces to Carol.

2 Carol trims at most two of the five pieces so as to create at least a three-way tie for largest. Setting the trimmings aside, she hands the five pieces – one or two of which may have been trimmed – to Ted.

3 Ted trims at most one of the five pieces he has been handed so as to create at least a two-way tie for largest. This, of course, may involve further trimming of a piece that Carol already trimmed in step 2. He sets his trimmings aside with those of Carol, handing the further altered collection of five pieces to Alice.

4 Alice now chooses from among the five pieces – some of which may have been trimmed by Carol and/or Ted – a piece that she considers to be at least tied for largest. (The remaining steps now reverse the order of initial play.)

5 Ted chooses next – from among the four remaining pieces – a piece that he considers to be at least tied for largest, with the proviso that if he trimmed a piece in step 3, and Alice did not choose this piece, then he must choose it now.

6 Carol chooses next – from among the three remaining pieces – a piece that she considers to be at least tied for largest, with the proviso that if she trimmed a piece or pieces in step 2, and one of these is still available, then she must now choose such a piece.

7 Bob now chooses – from the remaining two pieces – one that was not trimmed.

It is not hard to see that this procedure, and the foregoing player strategies, yield an envy-free allocation of *part* of the cake (i.e., the entire cake minus the trimmings and the one piece not chosen by any of the players). That is, because Alice chooses first, she certainly envies no one. Ted created a two-way tie for largest, and at least one of these two pieces is still available after Alice chooses, so he experiences no envy.

Carol, on the other hand, created a three-way tie for largest. If all of these pieces were unavailable to her, then it is because two were already chosen by Alice and Ted and the third was destroyed (i.e., made smaller) by Ted's subsequent trimming. But because of the proviso in step 6, the piece that Ted trimmed would definitely have been one of those chosen (by Ted or Alice). Hence, at least one of the three pieces tied for largest in Carol's eyes is available for selection by her.

Finally, Bob created a five-way tie for largest. At most two of these five pieces have been trimmed or chosen by Ted and Alice. Moreover, two may have been trimmed by Carol, and one chosen by her. This appears to represent a potential of five pieces trimmed or chosen, but the proviso in step 5 guarantees the two pieces trimmed by Carol cannot be disjoint from the set of pieces chosen: at least one of these trimmed pieces must already have been chosen by Carol, Ted, or Alice. Thus, a maximum of four pieces has been trimmed and/or chosen,[2] leaving at least one of the five original pieces untrimmed and unchosen and, therefore, available for Bob to choose.

In generalizing the $n = 4$ case to arbitrary n, Bob would begin by cutting the cake into $N = 2^{n-2} + 1$ pieces. As we just showed, for $n = 2$ we have $2^2 + 1 = 5$. Because N is an exponential function of n, it increases rapidly with increasing n: $n = 5$, 6, and 7 give $N = 9$, 17, and 33. As we shall show later, if we eliminate the provisos in steps 5 and 6 that a player must choose a piece that he or she trimmed if one is available, the required number of pieces that Bob must cut increases from 5 to 8 when $n = 4$, and very rapidly thereafter for larger n.

We can illustrate the trimming procedure in a manner similar to that which we did for the first stage of the Selfridge–Conway procedure for $n = 3$ (section 6.2). In the case of $n = 4$, however, what we do with the trimmings left over from steps 2 and 3, as well as the leftover fifth piece (perhaps now trimmed), is not so evident.

Suffice it to say at this point that we cannot disburse these leftovers, as we did with the trimmings in the case of $n = 3$, in a straightforward second stage so as to leave all players envy-free and the cake exhausted. In fact, as we will show later, the trimming procedure we use requires multiple stages to complete the process.

To illustrate the trimming procedure for $n = 4$, suppose the four players view the cake as having 20 units of value, with each unit illustrated by a small square, as before. Assume that each of the four players views the cake as follows:

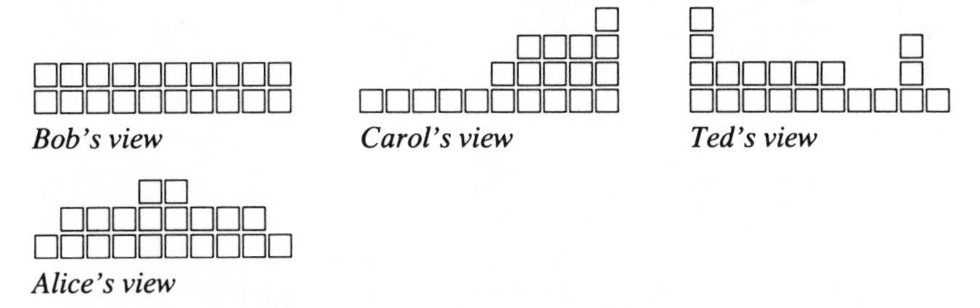

Bob's view *Carol's view* *Ted's view*

Alice's view

In step 1, Bob cuts the cake into five pieces he considers to be the same size, which from his viewpoint yields:

[2] In general, the total number of pieces trimmed and/or chosen by players n, $n - 1$, $n - 2$, $n - 3$, . . . is at most 1, 2, 4, 8, . . . , respectively. In the present situation, $n = 4$, so $n - 2$ (third term) corresponds to at most 4 pieces that have been trimmed and/or chosen.

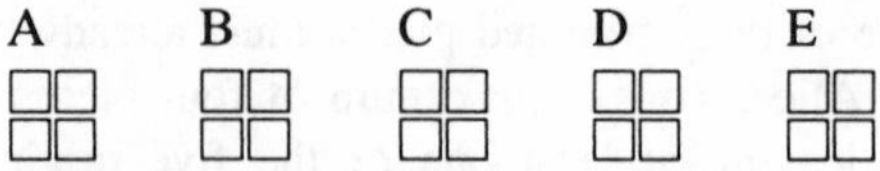

A B C D E

Bob's view of step 1

He now hands the five pieces to Carol, who views them as follows:

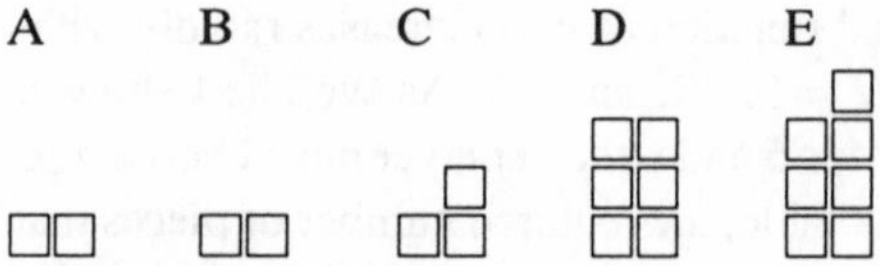

A B C D E

Carol's view of step 1

In step 2, Carol trims at most two of the pieces to create at least a three-way tie for largest. In our illustration, this means trimming D and E down to the size of C (i.e., 3 units of value), which can be achieved by trimming off the right-hand columns of both D and E:

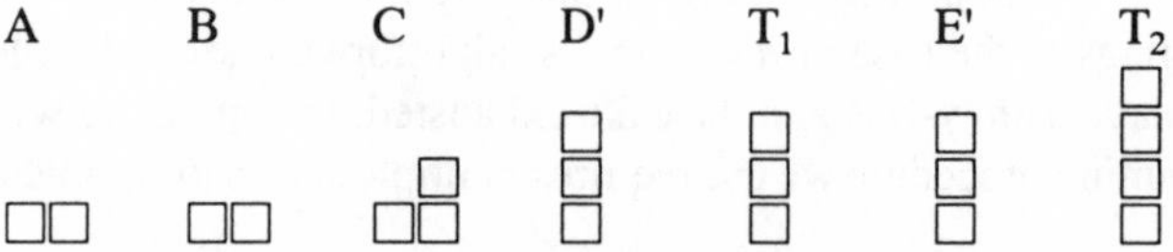

A B C D' T_1 E' T_2

Carol trims T_1 from D, yielding D', and T_2 from E, yielding E'

Notice that Carol now thinks C, D', and E' are all tied for largest (three units each).

After setting the trimmings (T_1 and T_2) aside, Carol next hands the five pieces (A, B, C, D', and E') to Ted, who views what he receives as follows:

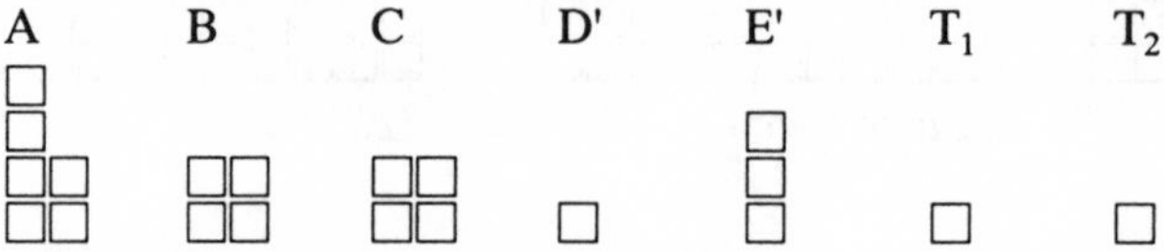

A B C D' E' T_1 T_2

Ted's view of step 2

In step 3, Ted trims at most one of the pieces to create at least a two-way tie for largest. This means trimming A down to the size of B and C (four units each):

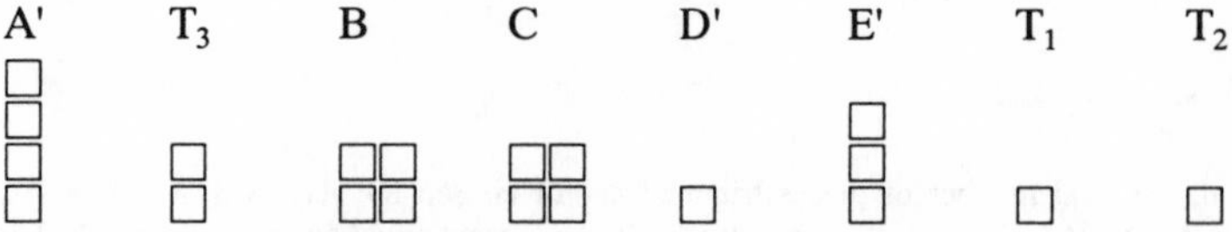

A' T_3 B C D' E' T_1 T_2

Ted trims T_3 from A, yielding A'

Notice that Ted now thinks A', B, and C, are all tied for largest (four units each).

Alice chooses first from among the five pieces (A', B, C, D', E) the one that she considers at least tied for largest. This will be C, which she thinks is 6 units (see first diagram).

Ted next chooses one of the two (or more) pieces he has tied for largest, subject to the proviso that if one that he trimmed is available, he must choose it. Because A' is tied for largest, he must choose it.

Carol chooses next, subject to the same kind of proviso as constrained Ted's choice. Thus, Carol must choose D' or E' (say, D', for definiteness).

Finally, Bob must choose a piece that was neither trimmed nor chosen. This will be B, the only available untrimmed piece.

Thus, starting with the entire cake and *n* players (four in our example), we have constructively obtained an envy-free allocation of part of the cake, leaving E and the trimmings comprising T_1, T_2, and T_3. Unfortunately, we cannot complete the process in a second stage, as with the Selfridge–Conway procedure for $n = 3$ (section 6.2).

If one allows for an infinite number of stages, however, the process is simple: one applies the procedure just illustrated over and over again, with each application yielding an envy-free allocation of the part left over from the previous application. Eventually (i.e., after infinitely many stages) the whole cake is allocated.

When the players all value a cake in exactly the same way, there will be no trimmings at the end of the first stage, so only one piece, which all the players agree is 1/5 of the cake, will be left over. At the other extreme, we can calculate the maximum amount of cake that might be left over after the first stage. Since Bob gets an untrimmed piece in the end, worth exactly 1/5 to him, he thinks the size of the leftover L_1 (all the trimmings plus the fifth piece remaining after all the players have chosen) is at most 4/5 of the original cake.

When we apply the procedure again to leftover L_1, then Bob will think the size of the leftover L_2 from the second stage is at most $(4/5)(4/5) = 0.64$ of the cake. To ensure that the maximum size of the leftover is less than 1/10 of the cake (in Bob's eyes) may take as many as eleven stages, because $(4/5)^{10} \cong 0.107$ and $(4/5)^{11} \cong 0.086$. On the other hand, if the players value the cake the same and there are consequently no trimmings, it will take only two stages to reduce the leftover to less than 1/10, because $1/5 = 0.20$ and $(1/5)^2 = 0.04$.

The procedure just described, which we call the *trimming procedure*, can be modified to eliminate the provisos in steps 5 and 6 that force Carol and Ted in the $n = 4$ case to choose trimmed pieces if they are available. This modification involves having Bob divide the cake into eight rather than five pieces that he considers equal. Next, Carol creates a four-way tie, and Ted creates a two-way tie, in steps 2 and 3, respectively.

The players would then make their choices of pieces, beginning with Alice, in the reverse order in which they cut or trimmed: Alice, Ted, Carol, Bob. The total numbers of pieces trimmed and chosen by the different players are:

Alice: 1 piece (0 trimmed and 1 chosen)
Ted: 2 pieces (1 trimmed and 1 chosen)
Carol: 4 pieces (3 trimmed and 1 chosen).

Now Alice gets a piece at least tied for largest, because she goes first. Ted also gets a tied piece, because he created a two-way tie for largest, and only one piece was chosen by Alice and none was trimmed. In addition, Carol gets a tied piece, because the number of pieces chosen and/or trimmed by Alice and Ted is at most $1 + 2 = 3$, and Carol created a four-way tie for largest. Finally, the number of pieces chosen and/or trimmed by Alice, Ted, and Carol is at most $1 + 2 + 4 = 7$. Because Bob created an eight-way tie for the largest, there is at least one unchosen and untrimmed piece remaining.

Note that the number of pieces trimmed and/or chosen is at least one less than the number tied for largest for the player about to choose, so there is always at least one largest piece remaining for that player. In general, a largest piece can be ensured for every player – whatever choice of largest piece (trimmed or untrimmed) that a player makes at each step – if Bob begins by cutting the cake into 2^{n-1} pieces. If $n = 5$, for example, then Bob would divide the cake into 16 pieces in order to ensure envy-freeness (rather than the nine required with the provisos to choose trimmed pieces if they are still available). Subsequently, Carol would create an eight-way tie, Ted a four-way tie, and Bob a two-way tie, and Alice would choose first.

Whether the trimmings are reduced rapidly or not, the process may take an infinite number of stages to complete.[3] Is it possible to complete the allocation in a finite number of stages and still ensure envy-freeness? The answer is yes, but it does not come so easily as applying the procedure again and again.

7.4 A finite procedure

To transform the infinite envy-free procedure for four or more people into a finite procedure, we pick up on an idea used in the Selfridge–Conway procedure for $n = 3$ (section 6.2).[4] Under the latter procedure, recall that Bob began the process by dividing the cake into three pieces he considered to be the

[3] For the set-theorists: Without any assumptions on the measures, if C is the cake and $|C| = \kappa$, then the cake will be completely allocated after a transfinite iteration of length κ.

[4] This section is adapted from Brams and Taylor (1995a) with permission; announcements of the result we describe have appeared in Olivastro (1992a), Gale (1993, p. 51), Campbell (1994, pp. 381–2), COMAP (1994), Hively (1995), and Stewart (1995).

same size. Then there was some trimming and selection of pieces that resulted in an allocation of part of the cake – that is, all except the trimmings T – among Bob, Carol, and Ted.

This partial allocation was not only envy-free but also one that Bob thought gave him a piece as big as Ted's (trimmed) piece *together with T*. Hence, Bob could rest assured that he would never envy Ted, regardless of how T was later distributed among the players. We referred to Bob as having an irrevocable advantage over Ted with respect to T: even if Ted ended up with all of T, Bob would believe he still had an equally large piece.

With this as background, we next describe a finite envy-free procedure for four people (it generalizes to any $n > 4$), based on the infinite procedure described in section 7.3. Assume for the moment the existence of an *IA subgame* (IA for "irrevocable advantage"), which we will say more about shortly. This subgame has the following property:

Suppose we are given a piece P of cake, as well as two slivers, which Bob thinks are the same size but Carol does not. Then the IA Subgame is played among all four players and results in an allocation of *part* of P (there is a leftover piece, L) among the four players that is envy-free. Moreover, Bob thinks he has an irrevocable advantage over Carol (with respect to L), and Carol thinks she has an irrevocable advantage over Bob (also with respect to L).

We ask the reader to suspend belief about the IA Subgame temporarily. We will call upon it as needed – without any explanation as to how or why it works – in order to describe the finite envy-free procedure, which works as follows:

1 Bob divides the cake into four pieces that he considers to be the same size. He then asks if any player objects because he or she thinks two of the pieces are, in fact, not the same size.
2 If no one objects, Bob hands out the pieces and we are done.
3 If, say, Carol objects, then we have all four players play the IA Subgame (which they can do, because Carol's objection ensures that we have the requisite two slivers, which in this instance are two of Bob's pieces) to obtain an envy-free allocation of all the cake except a leftover piece L. Moreover, no matter how L is eventually distributed among the players, neither Bob nor Carol will envy the other.
4 We now pretend that the previous leftover piece L is the whole cake, and we go through a slightly modified version of what we just described. (We will only have to modify what we did this one time; all future iterations will be exactly as what we are about to do.) The modification is the following:

Bob divides L (now the whole cake) into 12 pieces he considers equal, and we again ask if anyone objects.[5] (There is a reason for the number 12 that will

[5] If no one objects here, we are done. The reason we do not explicitly have to say this here is that it is covered in case 1 (next paragraph in the text).

become clear in case 1 below.) We separate the four people into two groups:

A, comprising those who agree the pieces are the same size; and

D, comprising those who disagree about their being the same size.

Notice that Bob is in the A group. We now consider two cases:

Case 1. Someone in the D group does not yet have an irrevocable advantage over everyone in the A group. (This is obviously true if L is the whole cake, which we will refer to as L_1, because no player has received anything yet.)

In this case, we choose such a player from the D group and a corresponding player from the A group. Then we have all four players play the IA Subgame, which we can do because we have the requisite two slivers, to obtain an envy-free allocation of all the cake except a new leftover piece, L_2. Moreover, we have now added one more pair of players to the list of those who have an irrevocable advantage over each other.

Although this completes what we do in case 1, it is worth pointing out that if we were to repeat what we are doing with L_2 in place of L_1, and then repeat this again with the new leftover piece L_3, and so on, case 1 can occur at most six times. This is because there are only six (unordered) pairs of people: Bob–Carol; Bob–Ted; Bob–Alice; Carol–Ted; Carol–Alice; Ted–Alice. Each time we go through case 1, we eliminate forever one more of these pairs as being the reason we will find ourselves in case 1 again. Thus, we will soon find ourselves in

Case 2. Everyone in the D group already has an irrevocable advantage over everyone in the A group.

Since Bob divided L into 12 pieces, we can now give the same number of pieces to everyone in the A group. (If A comprises two people, each gets six pieces; if A comprises three people, each gets four pieces; and if A comprises four people, each gets three pieces.) We claim this completes an envy-free allocation of all of the cake:

1 Certainly no one in A envies anyone else in A: the allocation up to this point has been envy-free, and everyone in A thinks the pieces just handed out are exactly the same size.
2 No one in A envies anyone in D: the players in D received nothing in this allocation of L.
3 No one in D envies anyone in A, which is the crucial fact. Because we are in case 2, *everyone* in D already had an irrevocable advantage over *everyone* in A.
4 Finally, no one in D envies anyone else in D, because the allocation so far has been envy-free, and nothing has changed for the players in D.

Before discussing the IA subgame itself, let us illustrate the sequence of steps we have just described with an example involving Bob, Carol, Ted, and Alice:

1 Bob divides the cake into four pieces, A, B, C, and D, that he considers to be the same size. He asks if anyone disagrees with his statement that they are all the same size.
2 In the scenario we postulate, Carol speaks up, saying that she thinks B and D are not the same size. (Let us assume that Ted also spoke up, whereas Alice did not. At this point, however, we only consider one such objection – the choice of Carol instead of Ted is arbitrary.)
3 All four players now play the IA subgame using the slivers B and D that Bob thinks are the same size and Carol does not. This yields an allocation W_1, X_1, Y_1, Z_1 (for Bob, Carol, Ted, and Alice, respectively) of all the cake except for a small leftover piece L_1. Moreover, it is an envy-free allocation for which Bob and Carol both think they have an irrevocable advantage over the other. (That is, Bob thinks that W_1 is at least as large as X_1 together with L_1, and Carol thinks that X_1 is at least as large as W_1 together with L_1.)
4 Bob now divides L_1 into 12 pieces that he thinks are the same size and again asks if anyone disagrees.
5 This time, Carol and Ted again say they disagree. Thus, we have the *A group* (consisting of those who agree the 12 pieces are the same size – Bob and Alice) and the *D group* (consisting of those who disagree – Carol and Ted). Notice that there is someone in the D group who does not yet have an irrevocable advantage over everyone in the A group; for example, Ted does not have an irrevocable advantage over Alice. (There are other examples we could work with equally well – Ted and Bob or Ted and Alice.) Hence, we are in case 1.
6 All four players play the IA subgame on L_1 using the pieces that Alice thinks are the same size and Ted does not. This yields an allocation W_2, X_2, Y_2, Z_2 (for Bob, Carol, Ted, and Alice, respectively) of all of L_1 except for a small leftover piece L_2. Moreover, it is again an envy-free allocation for which Alice and Ted both think they have an irrevocable advantage over the other. (That is, Alice thinks that Z_2 is at least as large as Y_2 together with L_2, and Ted thinks that Y_2 is at least as large as Z_2 together with L_2.)
7 Bob now divides L_2 into 12 pieces that he thinks are the same size and again asks if anyone disagrees.
8 This time, only Alice disagrees. Thus, the A group consists of Bob, Carol, and Ted, and the D group consists of Alice alone. Not everyone in the D group has an irrevocable advantage over everyone in the A group, because, for example, Alice does not have an irrevocable advantage over Carol.
9 All four players play the IA subgame on L_2 using the pieces that Carol thinks are the same size and Alice does not. This yields an allocation W_3, X_3, Y_3, Z_3 (for Bob, Carol, Ted, and Alice, respectively) of all of L_2

except for a small leftover piece L_3. Moreover, it is again an envy-free allocation for which Alice and Ted both think they have an irrevocable advantage over the other. (That is, Alice thinks that Z_2 is at least as large as Y_2 together with L_2, and Ted thinks that Y_2 is at least as large as Z_2 together with L_2.)

10 Bob now divides L_3 into 12 pieces that he thinks are the same size and again asks if anyone disagrees.

11 This time, Carol, Ted, and Alice all disagree. Thus, the A group consists of Bob alone, and the D group consists of the other three players. Not everyone in the D group has an irrevocable advantage over everyone in the A group, because, for example, Ted does not have an irrevocable advantage over Bob.

12 All four players play the IA subgame on L_3 using the pieces that Carol thinks are the same size and Alice does not. This yields an allocation W_4, X_4, Y_4, Z_4 (for Bob, Carol, Ted, and Alice, respectively) of all of L_3 except for a small leftover piece L_4. Moreover, it is again an envy-free allocation for which Bob and Ted both think they have an irrevocable advantage over the other.

13 Bob now divides L_4 into 12 pieces that he thinks are the same size and again asks if anyone disagrees.

14 This time Carol and Ted disagree. Thus, the A group consists of Bob and Alice, and the D group consists of Carol and Ted. But notice that we are now in case 2, since everyone in the D group has an irrevocable advantage over everyone in the A group. More precisely,
   - Carol has an irrevocable advantage over Bob from step 3;
   - Carol has an irrevocable advantage over Alice from step 9;
   - Ted has an irrevocable advantage over Bob from step 12;
   - Ted has an irrevocable advantage over Alice from step 6.

15 Bob now gives six of the 12 pieces into which he cut L_4 to Alice and keeps six for himself. (He and Alice comprise the A group.) Thus, the entire cake has now been allocated:
   - Bob receives W_1, W_2, W_3, W_4 and the 6 pieces from L_4;
   - Carol receives X_1, X_2, X_3, and X_4;
   - Ted receives Y_1, Y_2, Y_3, and Y_4;
   - Alice receives Z_1, Z_2, Z_3, Z_4 and the 6 pieces from L_4.

Certainly Bob does not envy Carol or Ted. Moreover, he does not envy Alice, since he thinks the six pieces from L_4 that he is getting are all the same size as the six pieces from L_4 that she is getting. Similarly, Alice envies no one.

Carol certainly does not envy Ted. Moreover, she had an irrevocable advantage over both Bob and Alice and so she envies neither of them. Similarly, Ted envies no one.

Thus, the allocation of the whole cake – in finitely many steps – is envy-free. This completes the description and illustration of the finite procedure (for the special case of $n = 4$, but the general case is conceptually no more difficult), subject to the existence of the IA Subgame.

Let us now say something about its existence: Given two pieces of cake, A and B, that Bob thinks are the same size, but Carol does not, we can arrive at six separate pieces of cake such that

- Bob thinks each of the first three pieces are the same size, and all are larger than each of the last three pieces; and
- Carol thinks each of the last three pieces are the same size, and all are larger than each of the first three pieces.

This is not hard to do, but we refer the reader to the appendix of this chapter for further details.

Given these six pieces, however, we can now complete the description of how the IA Subgame would be played. First, Ted trims at most one of the six pieces to create at least a two-way tie for largest. The players now choose among the pieces in the following order: Alice, Ted, Carol, Bob, with the stipulation that Ted must take the trimmed piece if it is available, and Carol must choose one from the second group of 3, whereas Bob must choose one from the first group of 3. Notice that Bob and Carol can do this, while still getting a piece that has not been altered by Ted or chosen by Ted or Alice. It is easy to see that this yields an allocation of part of the cake that is envy-free and, in addition, one that leaves Bob and Carol both feeling they have a strictly larger piece than the other.

All that is left now is to have the four players apply the sequence of steps from the infinite procedure over and over again until the remaining crumb is so small in the eyes of Bob and Carol that each thinks his or her advantage is irrevocable. This can be done because the crumb will become arbitrarily small, at least in the eyes of the player who does the initial division of the pieces being divided.

For our purposes here, therefore, we can simply let Bob and Carol take turns going first as we apply and reapply the trimming and choosing sequence from the infinite procedure. Eventually, when the size of the crumb left over is sufficiently small in the eyes of these two players, then neither will care if the other player should get the whole crumb.

7.5 Applying the trimming procedure to indivisible goods

While parceling out the last crumbs of a cake may occasionally pop up as a real problem (e.g., at a children's birthday party), the main practical problem in applying the trimming procedure is that many situations involve indivisible goods, which cannot be divided up at all, much less trimmed.[6] In chapter 4 we analyzed an envy-free procedure (AW) for allocating multiple goods between two players, which at most would require the splitting of one good (that on which an equitability adjustment had to be made). Although Knaster's procedure of sealed bids (sections 3.2 and 5.5) and Lucas's method of markers (section 3.3) would require no such splitting and are applicable to more than two players, neither is envy-free or equitable.

The trimming procedure, while not directly applicable to allocation problems involving discrete goods that cannot be split, can be adapted to such problems under certain conditions. The most important condition is that there be a sufficient quantity of more divisible goods, like small items or – even better – money, which can be trimmed in lieu of the discrete good.

Take, for example, the problem of dividing up an estate, in which a house is the single big item. Assume there are four (male) heirs, but only one thinks the house is worth more than 1/5 of the estate. If this person is the one to make the initial division, and if, in addition, he knows that the other heirs do not value the house so highly, he can begin by dividing up the estate into five pieces, with one piece being just the house. If none of the other three heirs thinks this indivisible piece has to be trimmed on the first round – even after other trimmings are made – then the house can, in effect, be "reserved" for the heir who thinks it is the most valuable piece.

This example illustrates how one player's knowledge of the preferences of the others need not always be exploitative but can, instead, facilitate the search for a solution. It will not always be apparent, however, precisely what information players should reveal and what they should hide (as is true in most negotiations). But because the trimming procedure has certain safeguards built in – in particular, allocating in stages in addition to ensuring envy-freeness in each – players probably can afford to be more open about their preferences than were these safeguards absent.

It is interesting to recall that when the allies agreed in 1944 to partition Germany into zones after World War II (first stage), they at first did not reach agreement about what to do with Berlin (section 2.5). Subsequently, they decided to partition Berlin itself into zones (second stage), even though this city fell 110 miles within the Soviet zone. Berlin was simply too valuable a "piece" for the Western allies (Great Britain, France, and the United States) to cede to the Soviets, which is at least suggestive of how, after a piece is trimmed off, it can be subsequently divided under the trimming procedure.

The idea of trimming to create ties also comes up in this case. After agreement was reached by Britain, the United States, and the Soviet Union on

[6] This section is adapted from COMAP (1994, chapter 13) with permission.

partitioning Germany into three zones (France was later given part of both the British and American zones), the United States and Britain discussed exchanging the two zones that they were scheduled to control. This did not happen in the end, but the United States received transit rights through the British zone to allay US fears of lack of access to the sea (Smith, 1963, pp. 16–17, 28–9), presumably making its zone as valuable as Britain's in the end.

The problem of what to do with Berlin was solved by dividing it into zones like the rest of Germany, but what if a large piece like Berlin is not divisible? In the settlement of an estate, this might be the house, as we suggested earlier, which may be worth half (or more) of the estate to the claimants. In this situation, there may be no alternative but to sell this big item and use the proceeds to make the remaining estate more liquid or, in our terms, "trimmable."

To illustrate the trimming procedure in the case of an estate, assume that the estate comprises six items. Four heirs have the following valuations for each item, which are indicated by points that sum to 100 for each:

	Heirs			
Item	Bob	Carol	Ted	Alice
1 House (H)	50	50	50	50
2 Boat (B)	20	10	10	10
3 Car (C)	10	20	10	10
4 Furniture (F)	10	10	10	10
5 Piano (P)	10	0	10	10
6 Art (A)	0	10	10	10
Sum	100	100	100	100

Notice the following features of this example: (1) all heirs consider H to be worth half the estate; (2) setting H aside, for Bob and Carol the most and least valuable items differ, whereas for Ted and Alice their valuations coincide yet differ from those of Bob and Carol.

Assume no heir has sufficient resources to pay off the other three to get H. Accordingly, the heirs agree to sell H on the open market. Suppose they get exactly 50 for it, which is what they agree it is worth. (If they get less – say, 40 – this would change their totals to 90, but this total would not affect the trimming process in a fundamental way.) After the sale, H becomes 50 (divisible) points rather than a single discrete item.[7]

[7] Besides conversion by sale, Young (1994, pp. 13, 134) discusses two other modes of division of indivisible goods: time-sharing (rotation); and dividing chances at getting the entire good (randomization). Although both these modes could also be expressed as 50 points – where each point stands for a certain share or a certain chance of winning – and the trimming procedure applied in the manner illustrated next, there is a problem with randomization (i.e., holding a

We start with Bob (it could be any of the heirs). He begins by dividing the estate (now items 2 through 6) into five parts, as prescribed by the trimming procedure. For him the following parts are all worth 20, which we underscore to indicate a tie for largest:

Bob: B C + 10 F + 10 P + 10 A + 20

If Carol goes next, she must create at least a three-way tie for largest by trimming from the parts that Bob created. Because she initially assigned 20 points to C and 10 points to A, C + 10 and A + 20 will be the largest parts for her (worth 30 each). Accordingly, Carol will trim each by 10 to create a tie with her next-largest item (F + 10), which is worth 20:

Carol: B C F + 10 P + 10 A + 10 (T = 20)

If Ted goes next, he must create at least a two-way tie for largest from Carol's parts. But because F + 10, P + 10, and A + 10 are all worth 20 to him, he need do no trimming, but his ties are different from Carol's:

Ted: B C F + 10 P + 10 A + 10 (T = 20)

Now Alice must choose a part which she considers to be at least tied for largest. This will be one of F + 10, P + 10, and A + 10, because her preferences are the same as Ted's. Assume she chooses A + 10, and Ted next chooses P + 10.

Then Carol has two remaining pieces, F + 10 and C, that she considers tied for largest. But since C is the result of her trimming a piece (namely, C + 10), Carol must take it (C) because it is available.

Finally, assume Bob chooses B. Then the first-stage allocation to heirs (Bob, Carol, Ted, and Alice) is (B, C, P + 10, A + 10), leaving F + 10, and the 20 trimmed by Carol, for the second stage of the procedure.

Now all the heirs value F equally (10), but none can divide F + 30 (F + 10 + 20) into five equal parts, as Bob did in the first stage of the procedure. Thus, we can see that there may be a problem of indivisibility at the second stage in some instances, which will necessitate selling other items to provide proceeds that can be divided in later stages.

In our example, however, there is no such necessity, because F + 30 can be exactly divided into *four* equal parts for the four heirs:

F 10 10 10

Adding these pieces, in this order, to the previous first-stage envy-free

lottery) that crops up after the procedure is applied. It is that someone will inevitably win the house, and therefore half the estate, making everyone else envious (even though each heir had the same shot before the winner was selected).

allocation, we obtain as a final envy-free allocation (B + F, C + 10, P + 20, A + 20) for (Bob, Carol, Ted, and Alice), as shown below:

Bob: B + F, giving 20 + 10 = 30 points
Carol: C + 10, giving 20 + 10 = 30 points
Ted: P + 20, giving 10 + 20 = 30 points
Alice: A + 20, giving 10 + 20 = 30 points.

Thus, our trimming procedure guarantees envy-freeness as long as there are enough divisible goods to ensure that no player ever has to trim a discrete good. Generally speaking, this means that there cannot be a single item, like a house, highly valued by all the players at the start, or a comparatively large item at a later stage.

If there is, such an item can be sold off to a nonplayer, as we assumed in our example. Alternatively, a player who desires it could pay off the other players in a negotiated settlement, which is simply another way of introducing more divisible goods into the system.

Still another way of lending divisibility to the trimming procedure is for all the players to make injections of cash. If, when combined with the indivisible goods, the cash gives enough "cushion" to each discrete good so as to make pieces trimmable when necessary, then it may be possible to accomplish the trimming without selling off any of the discrete goods.

In our example, this could be done if each heir contributed 150/4 = 37.5. Then Bob could divide the estate into five pieces, each worth 50:

Bob <u>H</u> <u>B + C + F + P + A</u> <u>50</u> <u>50</u> <u>50</u>

Since all the other players also value these pieces at 50 each, there would be no need for trimming in the first stage. One of the 50-pieces could be saved for the second stage and then immediately divided evenly among the four heirs.

Thus, there may be a solution to the problem of envy-freeness when there are discrete goods. Under the trimming procedure, they can be sold off when necessary to ensure sufficient liquidity, or each player need contribute only a relatively small amount to create sufficient divisibility of the items.

Of course, the infusion of cash under the trimming procedure is not unlike what the side payments under Knaster's procedure (sections 3.2 and 5.5) provide. But instead of liquidating the estate, or part of it, Knaster's procedure assumes that the players start with cash and then bid for items, with the relative winners of items compensating the relative losers so the losers end up not really being losers but in possession of at least their proportional shares. Unlike Knaster's procedure, however, the trimming procedure guarantees envy-freeness.[8]

[8] Fink (1995) has shown that, taking the point allocations as bids, Knaster's procedure happens to give an envy-free allocation in this example.

7.6 Efficiency, entitlements, and chores

1 Efficiency

We begin this section by returning to the question of efficiency that we analyzed for the proportionality procedures in sections 2.7 and 3.4. Part of the story for envy-free procedures has already been told. That is, in section 2.7 we indicated why procedures that are proportional (as are all envy-free procedures) are doomed to being inefficient. Consequently, neither the Selfridge–Conway discrete envy-free procedure in section 6.2 for $n = 3$, nor the discrete trimming procedure (finite version) in section 7.4 for $n > 3$, is efficient.[9]

For purposes of illustrating this inefficiency, consider the following application of the three-person Selfridge–Conway procedure to a cake comprising 1/3 vanilla (V), 1/3 chocolate (C), and 1/3 strawberry (S). Assume Bob likes each part equally, whereas Carol and Ted like V and C equally, each of which they prefer to S. Now consider three possible assignments according to the trimming procedure:

1 Bob divides the cake into three equal parts (for himself), each comprising V/3 + C/3 + S/3. Then Carol will do no trimming, because all pieces are identical, so these will be the pieces that go to each player.

2 Had Bob, possibly knowing Carol's and Ted's preferences, divided the cake into (V/2 + C/2, V/2 + C/2, S), then there would again have been no trimming. Consequently, Ted and Carol, in this order, would choose the two mixed pieces, and Bob would then receive S. Clearly, this allocation is better for Carol and Ted, and no worse for Bob, than assignment 1.

3 If Bob knows *how much* Carol and Ted prefer V and C over S, he can cut the cake initially to include a piece with all of S plus some of V and C for himself as well. As long as Carol and Ted consider this piece inferior to each half of an exact split of the remaining V and C – and do not feel spite for Bob's exploitation of them – they will choose the two halves of this split, and Bob will get all of S plus some of V and C as well.

Assignment 3 is strictly better for all three players than assignment 1, and better for Bob, but not for Carol and Ted, than assignment 2. In fact, both assignments 2 and 3 are efficient, but assignment 1, which the Selfridge–Conway procedure could give, is not.

[9] In a particular economic context, envy-free allocations that are also efficient may not even exist (Pazner and Schmeidler, 1974), necessitating tradeoffs among these and other properties (e.g., risk) that are explored in, among other places, Baumol (1986), LeGrand (1991), Fishburn and Sarin (1994a, 1994b), and Kolm (1995).

To illustrate the inefficiency of the trimming procedure with more than three players, consider a cake comprising three flavors, each of which is of equal volume and favored by one of three different players. Assume a fourth player, who is indifferent among the three flavors, divides the cake initially into five equal-size parts, each of which comprises an equal amount of the three flavors.

Then nothing will be trimmed, so this will be the allocation in the first stage, with the process being repeated over the division of the piece not chosen; in the final allocation, each player will end up with a mix of flavors. But this is worse for the first three players, and no better for the fourth player, than the allocation that gives each of the first three players 3/4 of his or her favorite flavor, with 1/4 of each of these three flavors going to the cutter, who is indifferent among them.

The envy-free moving-knife procedures also run amok of efficiency, primarily because they tend to preserve some form of contiguity in cake division. But if the best parts of a cake for a player are not connected, then satisfying contiguity may be inconsistent with satisfying efficiency.

Nuts, for example, are often distributed unevenly over the frosting of a cake, because the person doing the sprinkling is not usually as consistent as a salt spreader traveling at uniform speed. Therefore, the player who likes nuts will generally prefer different portions of the cake (where there is a heavy concentration), which may be disconnected. While this player could benefit from getting the little pieces with the heaviest concentration, the player allergic to nuts would have the opposite preferences.

None of the three-person moving-knife procedures will satisfy such players; they would prefer to cut up the cake into disconnected bits and pieces. This is especially true of the Stromquist procedure (section 6.3), in which the three slices that the players receive are contiguous. Under the Levmore–Cook (section 6.4), Webb (section 6.5), and the other two new moving-knife procedures (section 6.6), the pieces that a player receives may comprise two or more disconnected pieces, but each of these pieces is itself contiguous.

There is, however, a context in which a weaker form of efficiency is not only consistent with envy-freeness but actually implied by it (Berliant, Dunz and Thomson, 1992; Gale, 1993, p. 51).[10] A bit of terminology will help. Call a cake-division procedure for *n* players a *C-procedure* (for cut) if it results in

[10] Fleurbaey (1992), as cited in Arnsperger (1994, p.157), points out a different sense in which envy-freeness gives a weak version of efficiency – with respect to permutations, based on comparing allocations resulting from swapping whole portions among the players. Clearly, if swapping whole portions fails to make some players better off (envy-freeness), there cannot be an allocation, resulting from swaps, that is better for some players and not worse for others (efficiency).

a set of $n - 1$ cuts of a rectangular cake, all of which are parallel to the left and right edges of the cake.[11]

Thus, divide-and-choose can be regarded as a C-procedure for two players, as can the Dubins–Spanier moving-knife procedure for *n* people. Of more interest is the fact that Stromquist's moving-knife procedure for producing an envy-free allocation among three players is a C-procedure. On the other hand, the 3-person envy-free moving-knife procedure derived from Austin's procedure in section 6.5 is – at least as we described it – not a C-procedure since the allocation may involve three cuts (instead of two) for three players.

We say that a C-procedure for *n* players is *C-efficient* if there is no other C-procedure for *n* players that yields an allocation that is strictly better for at least one player and as good for all the others. With this terminology, we now have the following:

Proposition 7.1 An envy-free C-procedure is C-efficient.

Proof. The important thing to notice is that if one has two allocations arising from C-procedures for *n* players, then each allocation will have a piece that is a subset of the corresponding piece in the other allocation. For example, if there are three players (and thus two cuts), and if we have one set of cuts above the line that gives pieces (A, B, C), and one set below the line that gives pieces (A', B', C'), then we have the following:

A	B	C
A'	B'	C'

In this example, C' is a proper subset of C and A is a proper subset of A'. (The containments are both proper here, but we allow equality when we say "subset.")

Assume that the first allocation (A, B, C) is envy-free: the players who get A, B, and C – say, Bob, Carol, and Ted – all think that their piece is the largest or tied for the largest. Observing the above cuts, not only do all players know that C is larger than C', but because Bob and Carol think that A and B are at least as large as C, they also think that A and B are larger than C'. Therefore, all three players think that their pieces are larger than C', so whoever gets C' in the second allocation believes he or she did better under the envy-free allocation.

This means that the second allocation, (A', B', C), is not more efficient than the envy-free one, (A, B, C) – at least one player does worse under this

[11] The minimum number of cuts required for proportional and envy-free divisions was first raised by Steinhaus (1948, p. 103): "Interesting mathematical problems arise if we are to determine the minimal number of 'cuts' necessary for fair division." This question has been explored in Even and Paz (1984) and Robertson and Webb (1991; n.d.a).

allocation. Since the second allocation was arbitrary, every envy-free allocation (like the first) is efficient. Hence, there does not exist another one better for all players. Q.E.D.

Thus, we get C-efficiency "free" with envy-freeness for C-procedures. However, explicit in the notion of C-efficiency is that the allocation under one set of cuts, compared with another, is along a single dimension. In our example, both sets of cuts are along the horizontal dimension.

If we allow the comparison to be with respect to a second dimension (e.g., pieces cut horizontally are compared with pieces cut vertically), efficiency may not be preserved. Thus, consider a cake, as the players face it, whose lower third is vanilla, whose middle third is chocolate, and whose upper third is strawberry. Making vertical cuts that give each player 1/3 of the cake (in volume) obviously yields an envy-free allocation, because all pieces are identical in composition.

But this allocation is inefficient if each of the players prefers a different flavor. By making horizontal cuts, giving each player his or her favorite flavor, the allocation would be not only envy-free but also efficient. Thus, the argument that envy-freeness implies efficiency applies only if the $n - 1$ cuts are made along a single dimension, with the resulting pieces compared with other pieces also cut along this dimension.

Even cutting along a single dimension, most of the procedures we have analyzed require more than $n - 1$ cuts. Only Stromquist's three-person moving-knife procedure (section 6.3), in which there are four simultaneously moving knives but only two make cuts, meets the $(n - 1)$-cuts condition. Thus we have

> *Corollary 7.1 The Stromquist moving-knife procedure is a C-efficient C-procedure.*

Finally, we note that the approximate n-person moving-knife procedure would appear not to meet the $(n - 1)$-cuts condition sufficient to ensure it is a C-procedure, because players can reenter and call cut again and again. If, however, we count only the "final" cuts – assuming the substitutions erase the (temporary) cuts made earlier – the number of cuts will be $n - 1$ in the reconstituted cake. Hence, this procedure is "approximately C-efficient."

2 *Entitlements*

We turn now to entitlements, continuing the discussion begun in sections 2.8 and 3.4. There, however, the context was proportionality, where it was clear that entitlements could reasonably mirror greater proportional shares. At first blush, entitlements in the envy-free context seem contradictory. If you are entitled to twice as much as I am, then any reasonable allocation reflecting this two-to-one ratio will almost undoubtedly leave me envious of you.[12]

Surprisingly, it turns out, there is a natural way to marry the notion of entitlements and envy-freeness. The key to discovering this lies in Steinhaus' original method of handling entitlements in the proportional context.

Recall from section 2.7 that Steinhaus' idea was the following: If, say, Bob, Carol, and Ted are entitled to, respectively, 2/11, 4/11, and 5/11 of the cake, then an 11-person proportional procedure can (in this case) handle this by assuming that there are two players – that is, clones – with preferences identical to Bob's, four players with preferences identical to Carol's, and five players with preferences identical to Ted's. Each of Bob, Carol, and Ted then gets the pieces allocated to his or her clones by the 11-person procedure. The allocation is proportional in the sense of the given entitlements: Bob thinks he received at least 2/11, Carol thinks she received at least 4/11, and Ted thinks he received at least 5/11 of the cake.

Now what if we redo this same example, but using the *n*-person envy-free procedure from section 7.4 instead of using a proportional procedure? The resulting allocation will be envy-free: each of the 11 clones will experience no envy, thinking that he or she received at least 1/11 of the cake and no player received more. Combining clones, Bob will again think he has a piece of size at least 2/11, Carol will think she has a piece of size at least 4/11, and Ted will think he has a piece of size at least 5/11.

But more than this is true. The allocation to Bob, Carol, and Ted (via the 11 clones) is envy-free in the sense that there are partitions of Bob's piece into two subpieces, Carol's piece into four subpieces, and Ted's piece into five subpieces so that none of the three players would trade any one of his or her subpieces for one of the other's subpieces.

In general, if we have rational entitlements among *n* players, then we can use a common denominator q to express them as $x_1/q, x_2/q, \ldots, x_n/q$, where $x_1 + x_2 + \ldots + x_n = q$. (In our earlier example, $q = 11$, $x_1 = 2$, $x_2 = 4$, and $x_3 = 5$.) Given these entitlements, we will say that an allocation among the *n* players is envy-free if there exists a partition of the first piece into x_1 subpieces, the second piece into x_2 subpieces, and so on such that no player would gain by trading one of his or her subpieces for a subpiece held by another player.

Notice that this definition has two desirable properties:

1 If $q = n$ and $x_i = 1$ for every i (i.e., in the special case in which everyone is entitled to a proportional share of $1/n$), this definition of envy-freeness coincides with our earlier one.

[12] Presumably, this is why Young (1994, p. 12) said that envy-freeness "only applies when the parties have equal claims on the good."

2 An allocation that is envy-free in the context of entitlements is also proportional in the context of entitlements. (For example, suppose that an allocation is not proportional in the entitlements case because Bob thinks he received less than the 2/11 to which he is entitled. Then he would think that Carol and Ted have more than 9/11 between them. But then he would also think that at least one of the nine subpieces they hold is of size strictly greater than 1/11. Since Bob thinks his piece is smaller than 2/11, he must also think that at least one of his two subpieces is of size at most 1/11. Hence, Bob will want to trade, which shows that the allocation is not envy-free in the entitlements case.)

Both the n-person discrete and moving-knife procedures for producing allocations that are approximately envy-free can be adapted via Steinhaus' clone technique to handle entitlements as well. But because these extensions are straightforward, we will not provide further details here.

3 Chores

In section 2.8 we discussed the so-called chores problem – introduced by Gardner (1978) as the "dirty-work" problem – and proportional procedures for addressing it. Oskui (n.d.) gave three different envy-free chore allocation procedures for $n = 3$, two of which involve moving knives but none of which generalizes to a larger n.

Our approximate n-person moving-knife procedure (section 7.2) can be modified in two ways to provide an approximate envy-free solution to the n-person chores problem. Recall in section 2.8 that one way in which we modified the n-person proportional moving-knife procedure to handle chores was by having the first player to call cut receive the piece to the right of the moving knife and then exit the game. The last person to exit was then given the remaining cake.

To achieve approximate envy-freeness rather than proportionality, we add the rule that even if a person has already received a piece, he or she stays in the game and can continue to call cut. Should this person be the first to call cut again – because the new piece is ε smaller than his or her old piece – then this person will give back the old piece, which will be appended to the remaining cake, in return for the new piece.

The process continues in this manner until $n - 1$ pieces have been awarded, which must occur in a finite number of steps because of the ε tolerance level. At this point, the one person without a piece of cake will then be given all the cake that remains. He or she will consider this piece to be no larger by ε than any piece awarded earlier, because otherwise this person would have called cut earlier. Likewise, everybody else will consider his or her piece to be no larger by ε than any piece awarded later, because otherwise he or she would have called cut later. (These pieces cut later include the remainder of the cake awarded to the last player, which everybody will think cannot be smaller by ε than his or her own piece since everybody else received at least as large a piece earlier.) In this manner, our first modification of the n-person proportional procedure to handle chores can be adapted to give an approximately envy-free procedure for chore division.

Our second modification of the n-person proportional procedure in section 2.8 – in which the last person to call cut was awarded the piece to the left of the moving knife – can also be adapted to give an approximately envy-free procedure for chore division. As with the first modification, we assume that players do not exit the game once they have received a piece; in fact, they can continue to call cut for every new piece that comes up, which is substituted for their old piece if they are the last to call cut because they believe that the new piece is smaller than their old piece.

An analogous argument to that we just gave for the first modification of the n-person proportional procedure shows that the new procedure gives approximate envy-freeness for chores. As noted in section 2.8, the first and second modifications resemble Dutch and English auctions, respectively, but now with the possibility of bidding again.

Just as the approximate moving-knife procedure can be adapted to the chores problem, so can the trimming procedure (sections 7.3 and 7.4). Suppose, for example, that one must divide up chores such as mowing the lawn, going grocery shopping, and washing dishes. Assume the players have different preferences about those they least want to do, but everyone wants to do fewer chores rather than more.[13]

The trimming procedure is applicable to the chores problem, but it works in reverse: players "add on" rather than "trim" pieces. Like the trimming procedure, they do so by creating ties, though this time for the smallest rather than the largest piece.

A difficulty here is that players must first create a supply of the "bads" (instead of "goods"), or what we call a *pool*, to be used in the adding-on process, which players then draw from to create the requisite number of ties for smallest. They then choose from among these (tied) smallest pieces in a manner analogous to that of the trimming procedure.

But now it is the least of the bad rather than the most of the good that is allocated. As before, the procedure is multistage: although larger pieces always remain to be allocated after the smallest are assigned, the total left over still decreases after each round.

We illustrate the add-on procedure for the case of $n = 4$. Let Bob divide the

[13] If chores like this are not divisible, then they might be made so by counting the time spent on them, which is divisible.

chores into five pieces that he considers equal, and let Carol place tags on the two pieces that she considers largest (worst), or tied for largest. She then passes the tagged collection on to Ted, who tags the single piece so far untagged that he considers the largest or tied for largest. These three tagged pieces constitute the pool of chores from which Carol and Ted will draw to create ties for smallest.

This adding-on procedure, analogous to the trimming procedure, is then applied to the two remaining untagged pieces, with Carol and Ted this time creating ties for smallest by drawing from the pool. The fact that what bads each player contributed to the pool are bigger than the pieces for which each must create ties for smallest means that there is always enough "material" for each player to add on in order to create these ties.

Once the untagged pieces have been divided into envy-free portions, with possible leftovers, the procedure is then applied to these leftovers and the material remaining in the pool. We forego illustrating the add-on procedure with an example using little squares, like those we used earlier for cake, because such an application is quite straightforward. Suffice it to say that allocating bads in an envy-free manner, from household chores to civic duties like taxes (involuntary) and military service (voluntary), is probably as pervasive a problem as making envy-free allocations of goods.

In everything we have done so far, the challenge has been to overcome the difficulties caused by the players' having different preferences for whatever is being allocated. Achieving an allocation that everyone considers satisfactory (in some sense) is a chimera, however, unless there are procedures to implement it.

In chapter 8 we turn to a rather different aspect of the problem of fair division, in which what is fair is apparent but not how to induce the players to implement it. Suppose everyone values the goods to be allocated the same, but players have leeway as to how much of it they demand for themselves. Are there reasonable procedures – in the sense that they do not rely on unduly harsh punishments – that will result in players' making demands that yield a fair division? For these investigations, the cake is replaced by a dollar, so all proportional "pieces" must have the same value, and the game of divide-and-choose is replaced by a game called divide-the-dollar.

We will propose solutions in both the two-person and *n*-person cases. Although we give an example of how one of these solutions might be applied in a real-life situation, it is the ideas in this chapter rather than the application that we stress. Some of these ideas will be resurrected in chapter 9, where we propose a new auction procedure that, we believe, would be especially useful in situations in which there is incomplete information about the value of the objects being auctioned off.

Appendix

We prove here a supposition made in section 7.4: Given two pieces of cake, A and B, that Bob thinks are the same size, but Carol does not, we can arrive at six separate pieces of cake such that

- Bob thinks each of the first three pieces are the same size, and all are larger than each of the last three pieces; and
- Carol thinks each of the last three pieces are the same size, and all are larger than each of the first three pieces.

Assume Bob thinks A and B are the same size, and Carol thinks A is larger than B. Carol now names a positive integer $r \geq 10$, chosen so that, for any partition of A into r sets, Carol will prefer A, even with the seven smallest pieces – according to Carol – in the partition of A removed and placed with B.

Carol can easily choose such an r. That is, the union of the seven smallest pieces is certainly no larger than seven times the average size of all r pieces. Hence, Carol simply chooses r large enough so that $7\mu(A)/r < [\mu(A) - \mu(B)]/2$, where μ is her measure of the value of a piece.

Bob now partitions A into exactly r sets that he considers to be the same size, and he does the same to B.

Carol chooses the smallest three sets from the partition of B and names them Z_1, Z_2, Z_3. She also chooses either the largest three sets from the partition of A (which she does if she thinks these are all strictly larger than all the Zs), trimming at most two of these to the size of the smallest among the three, or she partitions the largest one of the sets in the partition of A into three pieces that she considers to be the same size. In either case, she names these Y_1, Y_2, Y_3.

Carol's strategy guarantees that she will think all three Ys are the same size, and strictly larger than all three Zs. This is true even if she chooses the second option.[14] Bob thinks all three Zs are the same size, and each is at least as large as all 3 Ys. This completes the proof.

[14] The proof runs as follows: We are assuming that both A and B have been partitioned into r pieces, and that B is not only smaller than A (in Carol's view) but smaller even than A with the smallest 7 pieces of A's partition removed. Arrange the sets in both partitions from largest to smallest (in Carol's view) as $A_1, A_2, \ldots, A_r$ and $B_1, B_2, \ldots, B_r$. Let μ denote Carol's measure, and suppose, for contradiction, that both of the following inequalities hold:

1. $\mu(B_3) \geq \mu(A_3)$, which holds if A_1, A_2, and A_3 are *not* all strictly larger than B_1, B_2, and B_3; and
2. $\mu(B_3) \geq \mu(A_1)/3$, which holds if A_1 *cannot* be partitioned into three sets all larger than B_3, B_2, and B_1.

It follows from inequality (1) that:

3. $\mu(B_7 \cup \ldots \cup B_{r-3}) \geq \mu(A_3 \cup \ldots \cup A_{r-7})$, since there are $r - 9$ sets in each union, and the smallest one of the Bs is at least as large as the largest one of the As.

It follows from inequality (2) that:

4 $\mu([B_1 \cup B_2 \cup B_3] \cup [B_4 \cup B_5 \cup B_6]) \geq \mu(A_1 \cup A_2)$, since each of the blocks of three Bs on the left-hand side is larger than each of the two As on the right-hand side.

But inequalities (3) and (4) clearly demonstrate that

5 $\mu(B) \geq \mu(A_1 \cup \ldots \cup A_{r-7})$.

This is the desired contradiction, because the set on the right-hand side of inequality (5) is A, with the smallest 7 pieces of its partition removed.

8 Divide-the-dollar

8.1 Introduction

Much work in the mathematical social sciences is devoted to showing the conditions under which individually rational actions can lead to collectively inferior outcomes (Kim, Roush, and Intriligator, 1992).[1] This problem is epitomized by the game of "Prisoners' Dilemma" (Brams, 1985a; Taylor, 1995), in which each player has a dominant, or unconditionally best, strategy of not cooperating, but the resulting outcome, and unique Nash equilibrium, is worse for both players than if they had both cooperated.

This clash between individual and collective interests is also illustrated by the game of *divide-the-dollar* (DD), wherein two players, Bob and Carol, independently propose a division of a dollar into cents, with each demanding a certain amount. Since we assume that Bob and Carol value the dollar, and parts of it, in the same way, the question is not how to carve out equal portions of it but, rather, how to induce them to do so on their own, with minimal sanctions for not being egalitarian (i.e., bidding 50 cents each). We stress "minimal," because we want egalitarian behavior, or something close to it, to emerge as a consequence of the players' rational calculations, not be imposed by an outside party or be the product of dire threats.

Thus, our search for procedures that result in allocations that players can implement themselves continues, but in a context in which the players are not faced with the problem of discovering a nonobvious allocation. In the less complex context of dividing a homogeneous good (a dollar) rather than a heterogeneous good (a cake), we seek a strategically compelling solution concept.

The usual rules of DD specify that each player receives whatever he or she bids if the sum of the bids does not exceed 100 cents; otherwise, the two players receive nothing. An intriguing feature of this game is that every ordered

[1] This chapter is adapted from Brams and Taylor (1994a) with permission.

pair (x, $100 - x$) for Bob and Carol, respectively, where $0 \leq x \leq 100$ – as well as (100, 100) – is a Nash equilibrium (defined in section 4.3).[2]

To illustrate, assume that the players propose (50, 50). If Bob raises his bid (say, to 51) or lowers it (say, to 49), he would do worse by receiving 0 in the first case and 49 in the second, assuming that Carol sticks with her bid of 50. Similarly, if the players propose (100, 100), each receives 0. But because neither can do better by raising or lowering his or her bid, (100, 100), like (50, 50), is stable in the sense of Nash.[3]

Unlike Prisoners' Dilemma, in which the players' dominant strategies lead to the inefficient Nash equilibrium, the problem in DD is that there are 102 pure-strategy Nash equilibria as well as many mixed-strategy equilibria (Myerson, 1991, p. 112).[4] In the absence of dominant strategies, it is difficult for the players to coordinate their bids in order to select one of the 101 (x, $100 - x$) efficient Nash equilibria – and avoid the (100, 100) inefficient Nash equilibrium – or nonequilibrium bids like (50, 60), that give the players payoffs of 0.

To be sure, only one of the plethora of efficient equilibria in DD gives the *egalitarian outcome* of 50 cents to each player, which is the unique symmetric Nash equilibrium in pure strategies that is also efficient. While this egalitarian outcome would seem the evident focal point of the players, however, there is little besides its "prominence," as Schelling (1960, pp. 56–8) puts it, to commend it as *the* rational choice.

Simple as DD is, this variable-sum game – in which the sum of the payoffs to the players varies between 0 and 100, depending on their bids – highlights the problem in which two players can both benefit if they make "reasonable" bids. However, if either player is too greedy, both may end up getting nothing.

[2] DD was introduced by Nash (1953) and is sometimes called the "demand game"; see van Damme (1991, pp. 145–50) for further analysis than is given here of its game-theoretic properties. In the related "ultimatum game," the proposals of the two players are not independent: Bob first makes a demand for a certain amount of money, and Carol responds. If she accepts his ultimatum, she gets the remainder; if not, both players get nothing. See Güth and Tietz (1990), Forsythe *et al.* (1994), and Gale, Binmore, and Samuelson (1995) for theoretical and experimental results on the ultimatum and related games. The ultimatum game can be extended to an alternating-offers game, whereby each player can make a new offer if he or she rejects the previous offer of the other player. With discounting, there is an efficient, "subgame-perfect" solution to this game (Rubinstein, 1982; Osborne and Rubinstein, 1990; 1994, chapter 3).

[3] Bids that each exceed 100, like (101, 101), also constitute a Nash equilibrium. But they are clearly less attractive than the other Nash equilibria, because each *guarantees* a player a payoff of 0 whatever the other player does. Consequently, we assume that no players would ever make bids above 100 in DD.

[4] *Mixed strategies* involve choosing, according to some probability distribution, different bids at different times, such as 50, say, 80 percent of the time, and 60 the remaining 20 percent of the time. All mixed-strategy equilibria are inefficient (van Damme, 1991, p. 146).

DD has been much discussed in the game-theory literature, but solving the coordination problem, much less justifying the egalitarian outcome, has proved difficult.[5] An attempt to rationalize salient choices by postulating conditions under which players, when confronted with multiple equilibria in a coordination game, reconceptualize their choices in a nonrational "involuntary phase," and solve a different (and more tractable) game in a "reasoning phase," is given in Bacharach (1993).

Our approach is different. We explicitly postulate new rules and show how they resolve the coordination problem by singling out the egalitarian outcome as uniquely rational.[6] Because the payoff scheme embodied in the new rules does not penalize the players for greediness to the degree that DD does, the egalitarian outcome can be obtained without the need to make incredible threats.

These rules, which describe variants of DD when there are two or more players, involve

- changing the payoff structure of DD to reward the lowest bidders first (DD1);
- adding a second stage that provides the players with new information yet restricts their choices at the same time (DD2); and
- both changing the payoff structure and adding a second stage (DD3).

We illustrate the possible applicability of the different procedures to a real-world allocation problem (setting of salaries by a team), in which there may be entitlements. We conclude by assessing the strengths and weaknesses of the different procedures and their solutions.

8.2 DD1: a reasonable payoff scheme

DD treats harshly bidders whose total request exceeds 100. Is it possible to render its payoff structure more reasonable, in some sense, and also to induce *egalitarian behavior* – bids of 50 by each player – as well as the egalitarian outcome of 50 cents to each player?

We use the term "reasonable" to rule out payoff schemes that induce egalitarian behavior by means of punishment for noncompliance to some preset standard. For example, a payoff scheme that gives each player 50 cents if he or she bids 50, and nothing otherwise, trivially induces egalitarian behavior. It accomplishes this feat, however, by dictating that some standard

[5] Philosophers, such as Lewis (1969) and Ullman-Margalit (1977), have argued that the coordination problem gets solved by the establishment of conventions or norms, but the rational foundations of this literature have been challenged (Gilbert, 1989, 1990).

[6] The consequences of another set of rules, which allows play to progress to a new round if the players make incompatible offers, is analyzed in Chatterjee and Samuelson (1990).

(i.e., bids of 50 by each player) must be met, lest the players get nothing,[7] which is the kind of Hobbesian solution we wish to rule out.

When a Leviathan steps in, the resulting coercive solution deprives people of the interesting and morally difficult choices they would otherwise face in real-life situations. The barbaric nature of such solutions also limits their practical use and effectiveness, especially insofar as they proscribe certain behaviors. Thus, the death penalty is a type of punishment that has questionable deterrent value, perhaps in part because it has been only fitfully administered; it is no longer used in many countries. In the United States, the prohibition of alcoholic beverages from 1920 to 1933 by the Eighteenth Amendment is an example of a ban that proved easy to violate and was revoked.

Call a payoff scheme in DD for n players *reasonable* if it satisfies the following five conditions:

1 Equal bids are treated equally.
2 No player receives more than what he or she bid.
3 If 100 units are sufficient to give every player what he or she bid, then these bids are the amounts disbursed to the players.
4 If 100 units are insufficient to give every player what he or she bid, then the 100 units are, nevertheless, completely disbursed to the players.
5 If all bids are greater than the egalitarian level of $100/n$, then the highest bidder does no better than the lowest bidder.

The question now becomes: Can one alter the payoff structure of DD in a reasonable way so that the egalitarian outcome is a solution, in some sense, in the corresponding game? (Note that the payoff scheme of DD violates condition 4.) Of course, "solution" could involve any one of a number of things, including dominant strategies, the iterated elimination of dominated strategies, and Nash equilibria.

In fact, the iterated elimination of *weakly dominated strategies* – strategies that are never better, and sometimes worse, than another strategy – is the concept we shall employ and illustrate shortly.[8] Such strategies always yield a Nash equilibrium. Like all reasonable schemes, this equilibrium does not result in egalitarian behavior, but it does lead to an egalitarian outcome:

[7] This payoff scheme is even harsher than DD, because it penalizes underbidding (i.e., bidding less than 50), whereas DD gives players at least their underbids, not 0, unless they bid 0. Clearly, there is a continuum of schemes that punish players for deviating, to varying degrees, from preset standards, but we do not consider them further for reasons given in the text. An experimental and theoretical literature that assumes players make bids according to the rules of DD, but have only incomplete information about the resources to be divided, is also worth noting; see, for example, Rapoport and Suleiman (1992) and references therein.

[8] This is the same solution concept used in Abreu and Matsushima (1994), who show in a different context how the severity of punishment can be mitigated by implementation via this concept. See also Jackson, Palfrey, and Srivastava (1994) and Sjöström (1994).

Theorem 8.1 Egalitarian behavior *is weakly dominated under every reasonable payoff scheme for n = 2. However, there is a reasonable payoff scheme, which works for all n, that yields the egalitarian* outcome *as the result of unique bids remaining after the iterated elimination of weakly dominated strategies.*

The proof of Theorem 8.1 is given in the appendix to this chapter. The proof of the second part of the theorem relies on showing that there is a reasonable payoff scheme that yields the egalitarian outcome as the result of unique bids remaining after the iterated elimination of weakly dominated strategies. Of particular interest here is the fact that the procedure works for more than two players.[9] The payoffs are made according to the following

Payoff Scheme (PS)

One starts with the lowest bidder – or, more generally, with the group tied for lowest bid – and pays them what they bid if there is enough money to do so. If there is not enough money, the money available is divided evenly among this group. One next moves to the group tied for second-lowest bid and proceeds in exactly the same way, but now working with only the money that is left after the group of lowest bidders has been paid. One continues in this fashion until the money is exhausted, after which no one else is rewarded, or all the players receive what they bid.[10]

Example

Suppose that four players (Bob, Carol, Ted, and Alice) bid 10, 60, 60, and 80, respectively. With PS, Bob would be paid first. He thus receives the full 10 he bid, and there is 90 remaining. Carol and Ted are tied for second-lowest bid (60) and are thus paid next. Since there is only 90 left (and not the 120 that

[9] Moulin and Shenker (1992) independently proposed a related scheme ("serial costsharing"). Showing that it uniquely satisfies certain properties (including coalition proofness), they applied it to cost sharing and surplus sharing. Their allocation mechanism specifies how much players share in the costs of developing a technology, based on how much they demand of its use, whereas we assume no distinction between inputs (costs of development) and outputs (benefits of technology). In our setup, the bids and payoffs are in the same currency (i.e., money), and doing well means getting a larger monetary payoff. By contrast, in the Moulin–Shenker scheme, doing well means paying lower costs in the development of a technology that will later be shared according to the players' demands. For further results on serial cost sharing, see Moulin (1994) and work cited therein.

[10] Demange's (1984) bidding procedure employs a similar payoff scheme, but her scheme becomes operative by default – only when a player objects to the highest bidder's proposed allocations to all players. Her procedure generalizes Crawford's (1979) two-person procedure to *n* players; see Young (1994, pp. 143–5) for further discussion.

would be needed to pay them both what they bid), each receives 45. The money has now been exhausted, so Alice, who bid 80, gets nothing. If, instead, the bids had been 40, 50, 80, and 80, the payoffs would have been 40, 50, 5, and 5, respectively.

We call DD having PS – instead of the usual payoff scheme – DD1, which defines a new procedure and hence a new game (i.e., with different rules of play). It is worth pointing out that there is still a multiplicity of equilibria under DD1. For example, egalitarian bids of 25 by the four players, or by only some players with the others bidding 26, are also Nash equilibria (see the appendix to this chapter). But it is easy to see that no bids other than 25 or 26 can constitute a Nash equilibrium, because either a higher bidder would always have an incentive to lower his or her bid, or a lower bidder would have an incentive to raise his or her bid.

Thus, all the other equilibria that PS generates are only slight perturbations of the Nash equilibrium of $b = 26$ by the four players, which reinforces the latter's robustness as the solution of DD1. Nonetheless, the fact that this solution requires 75 iterations to find (50 iterations would be required if there were only two players choosing $b = 51$, with the number of iterations approaching 100 as the number of players increases), is troublesome. We shall return to this issue after analyzing DD3, but next we describe and analyze DD2.

8.3 DD2: adding a second stage

Theorem 8.1 showed that there does not exist a reasonable payoff scheme for DD that yields egalitarian behavior, although DD1 comes very close in the sense of inducing equal bids that give the egalitarian outcome. This raises the second question we want to consider: Can one change the rules of DD so that the revised rules, together with a reasonable payoff scheme, yield egalitarian behavior?

Our starting point is a new game, DD2, which provides for a second stage but retains the old (unreasonable) payoff scheme of DD. We consider it for two reasons. First, it provides a simple context in which to introduce a new solution concept, "dominance inducibility," that we will need later. Second, the second stage, in which first-stage bids are revealed and can subsequently be revised, is pertinent to the analysis of two-stage auctions, which we will take up in chapter 9.

Definition

An outcome is *dominance inducible* by a player if that player has an opening move, or a choice, that – as the result of the successive elimination of weakly dominated strategies in the subgame induced by this move – yields that outcome.[11]

The rules of play of DD2, which is applicable only to two players, are as follows:

1 In stage I, players Bob and Carol make *initial* bids, x and y, which are made public.
2 In stage II, Bob and Carol make their *final* bids, b_1 and b_2, but the players are restricted in their choices to just x or y.
3 Based on the final bids in stage II, which are made public, the usual rules of DD determine the payoffs: if $b_1 + b_2 \leq 100$, Bob and Carol receive b_1 and b_2, respectively; otherwise, each receives 0.

Rule 2 may be interpreted as allowing a player to

- *affirm* its stage I bid; or
- *usurp* the other player's stage I bid.[12]

Of course, in stage I each player is free to choose any feasible stage I bid between 0 and 100. Consequently, the addition of stage II to DD enormously increases the number of possible strategies available to each player from 101 to 101 x 2^{100} because, for each of the 101 possible bids at stage I, each player may choose either his or her own bid, or the other player's bid, at stage II. (Because in exactly one of each of the 101 possible binary-choice situations at stage II the other player's bid will be the same as the first player's, each has 100 instead of 101 "my bid or your bid" decisions to make at stage II and hence 2^{100} *distinct* choices at this stage.) This explosion of strategies would appear to complicate play of DD greatly.

Surprisingly, just the opposite is the case: the addition of stage II singles out the egalitarian outcome as the only one that is dominance inducible by either player, as shown by

Theorem 8.2 In DD2 for two players, the egalitarian outcome is dominance inducible if either player bids 50. Moreover, it is the only outcome that is dominance inducible.

[11] Moulin (1979) called an outcome *dominance solvable* if it is the result of the successive elimination of weakly dominated strategies, which is true of the egalitarian outcome of DD1. When the elimination is with respect to strongly dominated strategies, the strategies that remain are said to be *rationalizable* (Bernheim, 1984; Pearce, 1984); see also Fudenburg and Tirole (1991, pp. 48–53). By contrast, dominance inducibility presupposes a prior choice by a player that makes the subsequent subgame – in particular, that played in stage II of DD2 – dominance solvable. As we shall show next, players in the stage II subgame always have dominant strategies, so the elimination of dominated strategies is immediate, not iterative.

[12] "Usurp" is not meant to imply that this bid is somehow taken away from the other player, who can affirm it (as his or her own) if that player so chooses.

Proof. Assume that Bob bids $x = 50$ and Carol bids y in stage I. We will speak of 50 and y as being *on the table* at this point because they are made public. The subgame describing what takes place once these bids are on the table, and therefore common knowledge, involves each player's independently choosing either 50 or y, with the payoffs as in DD. If $y = 50$, then "choose 50" is the only choice available (since there is only one number, 50, on the table at this point), and this choice dominates all others by default. The remaining two possibilities, $y > 50$ and $y < 50$, give the following payoff matrices:

Case 1. $y > 50$

		Carol		
		$x = 50$	$y > 50$	
Bob	$x = 50$	(50, 50)	(0, 0)	← Dominant strategy (weak)
	$y > 50$	(0, 0)	(0, 0)	
		↑ Dominant strategy (weak)		

Case 2. $y < 50$

		Carol		
		$x = 50$	$y < 50$	
Bob	$x = 50$	(50, 50)	(50, y)	← Dominant strategy (strong)
	$y < 50$	(y, 50)	(y, y)	
		↑ Dominant strategy (strong)		

The dominant strategies in stage II associated with $x = 50$ in these two cases demonstrate that the egalitarian outcome is dominance inducible by either player (Bob in the foregoing cases, who chooses $x = 50$ in stage I).

To see that no other outcome is dominance inducible by either player, note first that if a play of x_0 by Bob always leads to an outcome of (x_0, y_0) via the successive elimination of weakly dominated strategies, then we must have $x_0 = y_0$; otherwise, some stage I bids by Carol would preclude the outcome (x_0, y_0) from ever occurring, except when $x_0 = 0$, which is clearly dominated for Bob. Moreover, an outcome of (x_0, x_0) for $x_0 > 50$ is impossible. So it suffices to show that for $x_0 < 50$, the outcome (x_0, x_0) is not dominance inducible by, say, Bob.

To see this, consider the scenario wherein Bob bids $x_0 < 50$ at stage I and Carol bids $100 - x_0$ at stage I. Then, in the subgame describing stage II, Carol does strictly better by affirming $100 - x_0$ (when Bob affirms x_0)[13] than by usurping x_0. Thus, "choose x_0" is not the result of successive elimination of weakly dominated strategies in the subgame unless $x_0 = 50$. Q.E.D.

In fact, it turns out that if $x \neq 50$ and $y \neq 50$, there are four qualitatively different types of games that can occur, only two of which have a Nash equilibrium in pure strategies:

Game 1. $x > 50$ and $y > 50$.
The payoffs of the players are (0,0) at every outcome in the game matrix associated with stage II, ruling out these choices as part of a Nash equilibrium unless $x = 100$ and $y = 100$. If $x \neq 100$ and $y \neq 100$, then both players could have done better if Bob had bid $100 - y$, or Carol had bid $100 - x$, at stage I.

Game 2. $x > 50$ and $y < 50$ and $x + y \leq 100$.
This game is illustrated by $x = 60$ and $y = 40$. The payoff matrix for stage II is as follows:

		Carol	
		$x = 60$	$y = 40$
Bob	$x = 60$	(0, 0)	(60, 40)
	$y = 40$	(40, 60)	(40, 40)

Strategies associated with outcomes (40, 60) and (60, 40) are the pure-strategy Nash equilibria in this stage II game, which gives it some resemblance to the game of "Chicken."[14] But if, say, $y = 35$, then (35, 60) would not be a Nash equilibrium, because Bob could have done better by choosing a stage I bid of 40 and affirming it in stage II.

Game 3. $x > 50$ and $y < 50$ and $x + y > 100$.
The payoffs to the players are (0, 0) at every outcome in the stage II game matrix, except the (y, y) payoff associated with the players' joint choice of $y < 50$. But the strategies associated with this outcome do not constitute a Nash equilibrium, because Bob could have done better by unilaterally defecting to a strategy of bidding 50 at stage I and then affirming. This new strategy would

[13] If Bob usurps $100 - x_0$, they both get nothing.

[14] It would be Chicken if the preferences of the players were strict over the four outcomes, and the (40, 40) outcome in the matrix were next-best for both players, which would be the case by making it, say (45, 45) in the game 2 matrix. But the rules of DD2 do not allow the payoffs at this outcome to be different from the players' y bids.

give Bob a payoff of 50, as opposed to $y < 50$, whether Carol affirms $y < 50$ or usurps 50.

Game 4. $x < 50$ and $y < 50$.
The payoffs to the players are exactly what they bid at every outcome, but no strategies associated with these outcomes are in equilibrium because each player could have done better by raising his or her stage I bid.

In effect, the addition of stage II to DD, making DD2 a two-shot rather than a one-shot game, solves the coordination problem the players have under DD. Although (50, 50) is the prominent Nash equilibrium in DD, as we indicated before, it is not otherwise compelling. What Theorem 8.2 establishes is that *either* player can induce the rational choice of (50, 50) by making, at stage I, an initial bid of 50.

Note that this solution does not require that the players communicate with each other, much less make commitments or binding agreements, so play of DD2 is thoroughly "noncooperative" in the parlance of game theory. The communication, as it were, occurs when the stage I bids are made public and thereby become common knowledge (defined in section 4.2). Given that just one player bids 50 at this stage, the other player can do no better than also bid 50 at stage II.

There is a natural generalization of DD2 to the case where there are $n > 2$ players: all n players bid independently at stage I, and then each can, at stage II, choose to affirm his or her own bid or usurp any of the other bids. If the sum of the stage II bids is at most 100, then each player receives what he or she bid. Otherwise, all players receive payoffs of 0. But unlike the situation when $n = 2$, when $n > 2$ the solution given by Theorem 8.2 does not generalize, as shown by

Theorem 8.3 In DD2 for $n > 2$ players, no outcome is dominance inducible by any player.

Proof. As in the proof of Theorem 8.2, any outcome that is dominance inducible by a player must be of the form $(x_0, x_0, \ldots, x_0)$, where x_0 is the stage I bid of that player. Let y_0 be the bid corresponding to the egalitarian outcome. If x_0 were less than y_0, then a bid of y_0 by any other player and a choice by everyone to usurp y_0 would yield an outcome strictly better than $(x_0, x_0, \ldots, x_0)$ for everyone. If x_0 were greater than y_0, then the outcome $(x_0, x_0, \ldots, x_0)$ would be impossible. Finally, if $(x_0, x_0, \ldots, x_0)$ is the egalitarian outcome, then we need only consider the case in which Carol bids $x_0 - 1$ at stage I, Ted bids $x_0 + 1$ at stage I, and everyone else bids x_0 at stage I. That is, given these stage

I choices, it is now easy to construct scenarios at stage II in which (i) $x_0 + 1$ is a strictly better choice for Bob than is x_0; (ii) x_0 is a strictly better choice for Carol than is $x_0 - 1$; and (iii) $x_0 - 1$ is a strictly better choice for Ted than is $x_0 + 1$. Hence, "choose x_0 at stage II" is not the result of the successive elimination of weakly dominated strategies. Q.E.D.

Although DD2 does not employ a reasonable payoff scheme and breaks down when there are more than two players, the egalitarian outcome it induces does not require the successive elimination of many dominated strategies. Instead, this outcome can be induced immediately by a stage I bid of 50 by either player. The inclusion of stage II, which induces egalitarian behavior as well as an egalitarian outcome, gives DD2 an additional advantage: it makes the procedure *rectifiable* by enabling one player to "correct" his or her initial bid. Thus, if one player overbids or underbids and the other player bids 50 in stage I, the first player can, in effect, correct his or her mistake by usurping the bid of 50 by the other player in stage II.

8.4 DD3: combining DD1 and DD2

The third variant of DD we propose combines the idea of two stages from DD2 with the idea of successively paying off the lowest bidders from DD1. We call this variant DD3, whose rules of play we describe next.

In stage I, each of n players independently makes an initial bid. Once these bids are on the table and therefore known to all players, each player has a choice of affirming his or her own bid or usurping any of the other bids. (Thus, we allow for the possibility that all players choose to affirm or usurp the same bid in stage II.) Once the players choose their stage II bids, payoffs are made according to PS, exactly as in DD1. That is, one starts with those tied for lowest bid and proceeds to pay off as many bidders as one can until either the money is exhausted or all the players receive what they bid. Our main result is the following:

Theorem 8.4 In DD3 for $n \geq 2$ players, the egalitarian outcome is dominance inducible by any player. Only $n - 1$ successive eliminations of dominated strategies are necessary to arrive at bids leading to this outcome.

Proof. For simplicity, we shall again consider only the case $n = 4$ with egalitarian outcome (25, 25, 25, 25). Suppose that player 1 bids 26 at stage I. We will show that three eliminations of weakly dominated strategies yield "choose 26 from among the four or fewer bids on the table" as the sole remaining strategy choice for all four players at stage II. We first establish two claims:

Claim 1. A choice of 26 at stage II weakly dominates a choice of any $y < 26$ at stage II.

Proof. A choice of 26 guarantees a payoff of at least 25 to the player making this bid. Of course, a choice of y cannot yield a payoff greater than y, so 26 is strictly better than any bid $y < 25$, and at least as good as 25. However, if $y = 25$ and all the other bids are 0 and affirmed, then 26 is a strictly better stage II choice than 25.

Claim 2. Suppose that n (or fewer) stage I bids are on the table, one of which is 26. In addition, suppose that at stage II
(i) no player will choose any $y < 26$; and
(ii) no player will choose any bid among the k largest of those on the table.
Let y be the k+1st largest of the stage I bids. Then a choice of 26 weakly dominates a choice of y at stage II.

Proof. A choice of y, as described in the claim, can yield a payoff of at most 25, and this occurs when everyone else also chooses y. (That is, if anyone else chooses some $y' < y$, then a choice of y yields a payoff of at most $(100 - y')/3 \leq (100 - 26)/3 < 25$.) As before, a choice of 26 guarantees a payoff of at least 25. Thus, a choice of 26 is at least as good in every scenario as a choice of y. Moreover, if everyone else chooses 26, then a choice of 26 yields a strictly better payoff (26) than does a choice of y (≤ 22).

The theorem now follows immediately from claims 1 and 2. To begin with, the strategies 0, . . . , 25 are eliminated in the first reduction (by claim 1); the highest of the four stage I bids is eliminated in the second reduction (this is the $k = 0$ version of claim 2); the next-highest of the stage I bids is eliminated in the third reduction ($k = 1$ in claim 2); and, finally, the third-highest of the four stage I bids is eliminated in the fourth reduction ($k = 2$ in claim 2). The theorem said $n - 1$ reductions (instead of n reductions, as the four just described suggest), but it is easy to see that either the first or the last of the reductions is vacuous. (That is, if some bid is less than 26, then we cannot have three greater than 26.) Q.E.D.

DD3 would appear to provide the best of both possible worlds: a reasonable payoff scheme for n players, as in DD1; and the additional information given by stage II of DD2, which speeds up dominance inducibility to only $n - 1$ successive eliminations. But is stage II, an admitted complication, really necessary if DD1 also does the job, albeit in more steps? Before comparing the different solutions, we introduce an additional parameter (entitlements) and indicate how the procedures might actually be applied.

8.5 The solutions with entitlements

In this section we illustrate an application of some of the ideas discussed, but in a more general setting than that developed so far. Specifically, we assume that

- some players are entitled to more of the amount being divided than others; and
- these entitlements must be reflected in the solution (rather than its being an egalitarian apportionment).

Surprisingly, if there are only two players, each of whom has a different entitlement, DD2 fails to provide these, even when there is complete information about these entitlements. To illustrate this result, assume that the (male) president (P) of a company plans to award pay raises to two (female) vice presidents (VP1 and VP2), but he does not think they deserve the same raise. He announces that VP1, whom we call S, deserves only a small (s) raise, and VP2, whom we call B, deserves a big (b) raise.[15]

So as not to appear dictatorial, however, P says that he will not implement these raises but will instead ask each VP what she deserves. She may say medium (m) as well as s or b, but there are budgetary limits: whereas the company can afford one raise of s and one of b, or two of m, requests for more (i.e., one m and one b, or two b's) are unacceptable. In fact, consistent with the rules of DD2 as well as DD, P says that if the VPs make either of the latter pair of (unacceptable) requests, they will get no raises, making the payoff scheme unreasonable.

Despite P's announcement of which VP is more deserving, the unfavored S has no incentive to request only s. To see this, consider the following 3 x 3 payoff matrix, in which each player may bid s, m, or b in stage I and choose either her bid or the other player's bid in stage II (the payoffs at the conclusion of stage II are shown, whereas the strategies are the stage I bids):

[15] When players have different rights or claims, a variety of allocation rules have been explored using cooperative game theory; see, for example, O'Neill (1982), Aumann and Maschler (1985), Young (1987), Chun (1988), Curiel, Maschler, and Tijs (1988), Chun and Thomson (1992), Bossert (1993), Dagan and Volij (1993), and Fleurbaey (1994), in which solutions are derived from axioms. When different players (e.g., men and women searching for marriage partners, college students seeking roommates) have different preferences for each other, the problem of finding a stable matching is explored theoretically in Gusfield and Irving (1989) and Roth and Oliveira Sotomayor (1990); the latter work also contains an empirical analysis of the results of applying a matching algorithm to the placement of new physicians in hospitals for their internships and residencies (both the physicians and the hospitals can indicate their preferences for each other, but the hospitals' preferences take precedence, in a certain sense, giving them what might be called a greater entitlement in finding a stable matching).

		B		
		s	m	b
S	s	(s, s)	(m, m)	?
	m	(m, m)	(m, m)	(m, m)
	b	?	(m, m)	$(0, 0)$

These payoffs, except for the question marks, are an immediate consequence of what the players would bid (and receive) in stage II to maximize their payoffs. For example, when S bids s and B bids m in stage I, S can do no better than usurp m in stage II when B affirms m, resulting in (m, m).

As for the questions marks, if the players affirm their stage I bids in stage II, their payoffs will be (b, s) in the lower left cell and (s, b) in the upper right cell. But given P's announcement of the deservingness of S and B, the latter payoff assumption, which matches P's announcement, seems much more reasonable than the former. Hence, we assume payoffs of (s, b) in the upper right cell but leave open what the players' payoffs are in the lower left cell.

Even with this cell a question mark, S's strategy of m weakly dominates s. With s eliminated, B's strategy of m weakly dominates her strategy of b in the reduced 2 x 3 matrix, leaving a further reduced 2 x 2 matrix. Given the question mark, no further reductions are possible. Nevertheless, we conclude that neither player will bid what P announces to be her entitlement (s for S and b for B), because precisely these strategies are eliminated in the successive reductions. Hence, P's announcement about the deservingness of his VPs has no bite in inducing them to make the "right" (i.e., his proposed) choices.

To sum up, if there are entitlements and they are common knowledge, DD2 does not yield them in our example. DD1, on the other hand, is more successful:

Theorem 8.5 Suppose $e_1, \ldots, e_n$ are positive integers (entitlements) summing to k, and suppose that n players are allowed to submit bids (claims) $b_1, \ldots, b_n$. Define the greed of player i to be the number $g_i = b_i - e_i$. Assume under DD1 that players are paid off in the order determined by greed (least greedy to most greedy), with ties resulting in an allocation that is proportional to the entitlements of those involved in the tie. [Thus, if two players have entitlements of 8 and 12, and they bid 10 and 14 each, then they would divide a remaining pot of, say, 50 as (8/20)(50) = 20 for the first, and (12/20)(50) = 30 for the second.] Then iterated domination of weakly dominated strategies results in a bid of $e_i + 1$ by player i for each i.

Sketch of proof. Arguments similar to those in claims 1–8 of the proof of Theorem 8.1 (see the appendix to this chapter) show that any bid $x \leq e_i$ by player i is weakly dominated by a bid of $e_i + 1$. Now assume that some bid $x > e_i + 1$ remains after the iterated elimination of weakly dominated strategies. Among all such x's (for all players), choose one making $g_i = x - e_i$ as large as possible. Thus, a bid of x will leave player i as most greedy, or perhaps tied for most greedy. It then follows easily that a bid of x yields player i a payoff of at most e_i. Thus, a bid of $e_i + 1$ in place of x by player i would never be worse, and it would be strictly better if everyone else bid one unit more than their entitlement. Q.E.D.

As with the egalitarian outcome without entitlements under DD1, no player receives his or her bid of $e_i + 1$ but instead the exact entitlement of e_i. Thus, optimal bids are perturbations of the egalitarian outcome, but they are only slight, rendering outrageous requests under DD1 nonoptimal.

As a possible application of DD1 (or DD3), consider a team of players that works closely together (e.g., in a company or on the athletic field). Assume that the team must set raises for its members, based on their previous performance, that are to be taken from a preset pool of money.

Each player (assume they are all female) may request any amount for herself up to the size of the pool. At the same time, each player makes a recommendation of a pay raise for every *other* player, with the sum of her own request and her recommendations for all the other players equal to the pool. The recommendations for each player – by every other player – are averaged to determine each player's entitlement.

DD1 is then applied, with allocations made in the order of the closest matches between the entitlements and the requests of each player. Given that the players make honest assessments of their teammates, then they have an incentive to make honest assessments of themselves.

To see why, assume that everybody requested the entire pool for herself. This is obviously not a Nash equilibrium, because every player would then have an incentive to lower her request slightly and thereby receive almost the entire pool, leaving almost nothing for anybody else. This logic eventually will carry the players towards the ratings they think the other players will give them. Provided players are honest in their assessments of others (there seems no good reason why they should not be in the absence of collusion), players can do no better than try to reflect the others' assessments, slightly perturbed, *in their own requests*.

Of course, incomplete information may prevent a perfect match. Nevertheless, we believe DD1 or DD3 would be viewed as fair by the players, because these procedures benefit players whose self-ratings agree with those of others. If there is exaggeration and posturing, it would more likely come in bargaining over the size of the pool available for salary raises rather than in the individual requests by the players.

8.6 Conclusions

Divide-the-dollar (DD), which punishes players by giving them nothing if their bids exceed 100, has a multiplicity of Nash equilibria. Although the egalitarian outcome is prominent, it is not otherwise distinguishable in noncooperative play. Furthermore, DD does not satisfy five conditions of reasonableness that preclude punitive behavior, which is often difficult to enforce even if severe punishment is "on the books." Practically speaking, why threaten such punishment if the threat is likely to be empty and there are, moreover, "softer" ways of inducing reasonable behavior?

We showed that an alteration in the payoff structure of DD, whereby the players who bid the least are paid off first (DD1), can induce the egalitarian outcome via iterated elimination of weakly dominated strategies. Like all payoff schemes that satisfy the reasonableness conditions, however, the players' optimal bids are not egalitarian but slightly greater.

If the rules of DD are revised to add a second stage but leave intact the old payoff structure, the resulting procedure is DD2. We showed that, based on dominance inducibility, it gives both the egalitarian outcome and egalitarian behavior if there are only two players, but it fails if there are more players.

Combining the second stage of DD2 with the payoff scheme of DD1 gives DD3. Under this procedure, which, like DD1, is reasonable, the successive elimination of weakly dominated strategies is greatly speeded up, which makes its solution more transparent than that of DD2. Like DD2, it can be implemented if only one player makes an egalitarian bid in stage I because of dominance inducibility; and it is rectifiable, because a player who (mistakenly) does not make an egalitarian bid in stage I can usurp the egalitarian bid of another player in stage II.

In addition to not generalizing to $n > 2$, DD2 does not hold up well when entitlements are introduced, even for two players, whereas DD1 – and by extension, DD3 – does. Provided the players on a team are sincere in evaluating each other team member's merits, DD1 and DD3 encourage honest assessments of self-worth, only slightly perturbed upward.

The main advantage that these procedures have over letting others, such as a boss or teammates, be the sole determinants of one's salary increase is that they encourage personal responsibility. One cannot simply blame others for a faulty evaluation if one's own estimate partially determines the result.

Both DD1 and DD3 incorporate in their payoff functions a person's request in such a way as to reward him or her for a searching and accurate self-assessment. This makes it rational for a person to gather information about others' perceptions of his or her performance before making a request. By adjusting one's request to others' perceptions, DD1 and DD3 induce one to see oneself as others see one. Psychologically speaking, this is probably good not only for fostering more realistic attitudes but also for promoting better team performance.

Appendix

Theorem 8.1 Egalitarian behavior *is weakly dominated under every reasonable payoff scheme for $n = 2$. However, there is a reasonable payoff scheme, which works for all n, that yields the egalitarian* outcome *as the result of unique bids remaining after the iterated elimination of weakly dominated strategies.*

Proof. To prove the first part of Theorem 8.1, it suffices to show that, in a two-person game, if the payoff scheme satisfies the five conditions of reasonableness, then a bid of 51 weakly dominates a bid of 50. To show this, we divide an opponent's bids into four cases and demonstrate that, in all cases, a bid of 51 yields at least as good an outcome as does a bid of 50 and, in at least one case, a strictly better outcome.

Case 1. Opponent bids $b \leq 49$.

In this case, a bid of 51 yields an outcome of 51 by condition 3. By the same condition, a bid of 50 yields an outcome of 50, so a bid of 51 is strictly better.

Case 2. Opponent bids $b = 50$.

In this case, a bid of 51 yields an outcome of at least 50 by conditions 2 and 4. On the other hand, a bid of 50 cannot yield a better outcome by condition 2 again.

Case 3. Opponent bids $b = 51$.

In this case, a bid of 51 yields exactly 50 by conditions 1 and 4. As before, a bid of 50 can yield no more than 50 by condition 2.

Case 4. Opponent bids $b \geq 52$.

In this case, a bid of 51 yields an outcome of at least 50 by conditions 4 and 5. A bid of 50 cannot yield a better outcome by condition 2. This completes the proof of the first part of Theorem 8.1.

To prove the second part of Theorem 8.1, we will show that there is a reasonable payoff scheme that yields the egalitarian outcome as the result of unique bids remaining after the iterated elimination of weakly dominated strategies. The payoffs are made according to the following payoff scheme (given in section 8.2):

Payoff Scheme (PS). One starts with the lowest bidder – or, more generally, with the group tied for lowest bid – and pays them what they bid if there is enough money to do so. If there is not enough money, the money available is divided evenly among this group. One now moves to the group tied for second-lowest bid and proceeds in exactly the same way, but now working with only the money that is left after the group of lowest bidders has been paid. One continues in this fashion until the money is exhausted, after which no one else is rewarded, or all the players receive what they bid.

It is easy to check that PS satisfies the five conditions of reasonableness given in section 8.2. Our claim now is that the iterated elimination of weakly dominated strategies results in unique bids that yield the egalitarian outcome.

For clarity of exposition, we will work with the case $n = 4$ and show that the outcome (25, 25, 25, 25) results from simultaneous bids of 26 by all four players. Moreover, these bids are the only ones left after a sequence of 75 successive eliminations of weakly dominated strategies. The arguments all extend to the case where there are n players and kn units ($n, k \geq 2$), where k ("cents" in DD) is the egalitarian bid of each player (25 in our example). The result we now prove is an immediate consequence of the following ten claims:

Claim 1. A player never receives more than what he or she bid.

Proof. This is obvious from the description of PS.

Claim 2. Suppose that in making the payoffs, there are exactly t players not yet paid. Suppose that Bob is one of these, and none of the other $t - 1$ has a lower bid than Bob. Then Bob receives at least the minimum of two quantities – what he bid, or $1/t$ of the money that is left over.

Proof. If Bob's bid is the lowest of those remaining, he receives either this bid or, if this bid is greater than the amount of money that is left over, the money that is left over. If Bob's bid ties for lowest with one or more of the remaining bids, the tied players receive either their bids or, if the sum is greater than the amount that is left over, they split this amount. Thus, the claim is true whether or not there is a tie.

Claim 3. A bid of 26 guarantees a payoff of at least 25.

Proof. Let s denote the number of players bidding less than 26. Then $4 - s$ bid 26 or more. By claim 1, at most $25s$ cents are needed to pay off the s players bidding less than 26. Thus, at least $100 - 25s$ cents are left over. By claim 2, Bob receives at least the minimum of 26 and $[1/(4–s)]\ [100 - 25s] = 25$.

Claim 4. A bid of $x \leq 25$ is never better than a bid of 26.

Proof. This follows from claims 1 and 3.

Claim 5. If $x \leq 25$, then there are scenarios in which a bid of 26 is strictly better than a bid of x.

Proof. If everyone else bids 0, then a bid of 26 yields a payoff of 26, whereas a bid of x yields a payoff of x.

Claim 6. For $0 \leq x \leq 25$, a bid of 26 weakly dominates a bid of x.

Proof. This follows from claims 4 and 5 – that is, a bid of 26 is at least as good as, and sometimes better than, a bid of 25 or less.

Claim 7. For $26 \leq x < y \leq 100$, neither x nor y weakly dominates the other.

Proof. If Bob bids x, and the other three players also bid x, then these bids yield payoffs of 25 to everyone. If Bob bids y, this yields him a payoff of at most 22 (since each of the other three will receive at least 26). Thus, we have a scenario wherein x is a strictly better bid than y. For a scenario in which y is strictly better than x, consider the situation in which Bob bids y, Carol bids x, and everyone else bids 0. Then x yields x and y yields y when $x + y \leq 100$.

Claim 8. In the first reduction caused by each player's elimination of dominated strategies, the strategies $0, 1, \ldots, 25$ are precisely the ones that are eliminated.

Proof. This follows from claims 6 and 7.

Claim 9. Assume that $26, \ldots, j$ are the only strategies that have not yet been eliminated, where $27 \leq j \leq 100$. Then j is weakly dominated by x for every x such that $26 \leq x < j$.

Proof. The only way j could be better than x is if the sum of the other three bids were less than $100 - x$. (Thus, more than x is available, so a bid of "x" yields x and a bid of "j" yields more than x.) But $x \geq 26$, so $100 - x \leq 74$, whereas the sum of the other three bids is at least 78. Thus, x is at least as good as j in every scenario. Moreover, if everyone else bids x, then x yields a strictly better outcome (25) than does j (≤ 22).

Claim 10. With the same assumptions as in claim 9, if we have $26 \leq x \leq y \leq j - 1$, then neither x nor y weakly dominates the other.

Proof. As in the proof of claim 7, it is easy to see that x produces a better outcome if everyone else bids x, whereas y produces a better outcome if everyone else bids j. Q.E.D.

9 Fair division by auctions

9.1 Introduction

What do auctions have to do with fair division? In this chapter we will show that bidding for items, and awarding them to the highest bidder – or, possibly, splitting the award among two or more of the highest bidders – may provide a compelling means to allocate them.

There are a plethora of different auction procedures, including English auctions, Dutch auctions, sealed-bid auctions, Vickrey auctions, and many others. We have already alluded to English and Dutch auctions in connection with the last-diminisher procedure (section 2.3) and n-person proportional and envy-free procedures (sections 2.7 and 7.6). We also analyzed Knaster's procedure of sealed bids (section 3.2), comparing it with the AW procedure, in which players allocate points as if they were bidding (section 5.5).

The auction procedure we analyze in this chapter, in which bidding is carried out in two stages, is unorthodox. It has not heretofore been described or analyzed in any publications.[1] We shall give reasons why we think this auction procedure is, in many situations, superior to others, some of which we shall discuss. But here we simply note that our analysis rests partly on ideas we developed in chapter 8, including that of adding a second stage, and having players successively eliminate weakly dominated strategies, in divide-the-dollar.

In a two-stage auction, players submit sealed bids for an item in a first stage. These bids are then opened and made public, but the identity of the players who made them is not revealed.

The players then have the opportunity to bid again in a second stage, but they are restricted in their choices: they can choose only from the bids already made in the first stage (including their own). The highest bidder at the second stage wins the item, unless there is a tie for highest, in which case it is broken in favor of the player (among those tied) who made the highest first-stage bid.

[1] Our analysis is based on two preprints (Brams and Taylor, 1994d, 1994e), which contain other results not included here.

In the context of fair division, two-stage auctions have several attractive features. Chief among these is that they provide players with the opportunity to revise their bids in light of information they obtain after the first stage. This feedback gives players a more informed basis for assessing the worth of an item and, subsequently, bidding intelligently in the second stage, when their bids determine the outcome (except in the case of ties, when first-stage bids also count).

We view such auctions as fairer than other kinds of auctions because players are less prone to make mistakes in them – or, if they make mistakes in the first stage, they may be able to rectify them in the second stage.[2] Furthermore, because bidding tends to be honest or sincere in two-stage auctions, as we will show, players win items that they genuinely value more than other players, which makes them, in a sense, more deserving of them. Furthermore, insofar as players have similar resources but prefer different things, they will get those things they most deserve, leading to a fair distribution of rewards to different players.[3]

In section 9.2 we compare two-stage auctions with "Vickrey (second-price) auctions." Like Vickrey auctions, in which the highest bidder wins but pays only the second-highest bid, two-stage auctions encourage players to be honest or sincere in their bidding. This is especially true in the first stage of "private-value auctions," in which a player's valuation of an item is assumed not to be influenced by that of other players.

Two-stage auctions, as we show in section 9.3, are superior, in certain ways, to Vickrey auctions. For example, players may "bail out" in two-stage auctions, which seems especially important if they learn, in the first stage, that their valuation of an item is far greater than that of other players. In addition, two-stage auctions facilitate the identification of shills or confederates, whom an auctioneer may introduce into a Vickrey auction to raise the price (by pegging it just below the highest bid) that the winner must pay. Finally, two-stage auctions better allow for "split awards" – in which a divisible good is apportioned among two or more bidders – than one-stage auctions, for reasons we discuss in section 9.5.

It is in "common-value auctions" – in which players' valuations are influenced by what happened in the first stage as they strive to ascertain the true worth, or common value, of an item – that the second stage is most valuable. It provides a means to avoid the "winner's curse" in one-stage auctions, whereby

[2] We shall say what we mean by "mistakes" later.

[3] To be sure, if the players are not in similar financial circumstances, this may not be the case. In the vernacular of economics, the fairness of two-stage auctions – in distributing items to those most deserving of them – can be challenged when players do not have similar endowments, which are the resources with which they start out.

being the winner may mean that one overpays for the items one wins. Although one may discount bids in one-stage auctions to ameliorate this problem, as we show in section 9.4, it is more efficient to use first-stage bids, especially if they are sincere, to make a best estimate of common value before bidding in the second stage.

Whether bidders use the median, mean, or some other measure of first-stage bids as the best estimate of common value, there may be no equilibrium strategy, as we show in section 9.5. But it is this very lack of stability, which gives the players few guidelines for making strategic bids, that may induce them to be sincere in the first stage. Once they know the distribution of (presumably sincere) first-stage bids, they will be in a better position to avoid the winner's curse.

Although auctions may seem a more unusual procedure to settle fair-division questions than others we have discussed, they are extremely common, used to sell everything from stamps to oil tracts to airwaves. Generally speaking, when (1) there is an item that can be sold without complex negotiations or intermediate market makers, (2) one does not know exactly what price to ask for it, and (3) one wants to ensure that it is available to all who might be interested in it through a competitive process, an auction is often a good way to sell it.

An outcome in an auction is *efficient* if the bidder who places the highest valuation on an item makes the highest bid and, therefore, wins it.[4] If different players accord the highest valuation to different items either in the same auction or different auctions, a fair auction procedure will result in a distribution of awards to the most deserving players, at least if the players are similarly endowed.

As just one example of this, consider the case of a university in a town with several roofing contractors. Every year, the university repairs or replaces many roofs, and the work is always put out to bid. If over time different contractors win different contracts, this will help keep most of the contractors in business and bidding, therefore, more competitive.

After summarizing the main features of two-stage auctions in section 9.6, we consider the prospects for their adoption. Support for their adoption will depend critically on how the bid taker – who is, after all, the person who usually sets the rules – is likely to fare under them. We argue that the propensity of two-stage auctions to induce sincere bidding may well override any temporary advantage that the bid taker might derive from the winner's curse, which is now well known and, therefore, unlikely to benefit the bid taker over an extended period of time.

[4] When this is not the case, the highest-valuation player can be made better off by awarding him or her the item and having that player pay the winner what the winner bid (making the winner no worse off, at least if his or her bid was sincere).

9.2 Two-stage auctions

In a review article on auctions, McAfee and McMillan (1987, p. 711) remark that "the essence of the auction problem is the unobservability of bidders' valuations." We concur. To mitigate this problem, we propose the following rules of play for a *two-stage auction*:[5]

1 *Stage I.* The players submit sealed bids, all of which are then opened and made public. They have no prior information about each other's sealed bids or valuations.[6]
2 *Stage II.* Each player chooses exactly one of the stage I bids, either by affirming his or her own bid or by *usurping* another player's bid.
3 *Payoffs.* If there is *only one* player who makes the highest bid in stage II, that player wins the auction, regardless of what he or she bid in stage I. If there is *more than one* player who makes the highest bid in stage II, the player who made the highest bid in stage I wins the auction.[7]

In terms of the information they provide, these rules give two-stage auctions elements of both an English auction, in which bids are oral, open, and ascending, and a sealed-bid auction. Unlike an English auction, bids are revealed all at once (in stage II), not sequentially; but like an English auction, bids can be revised. Unlike a sealed-bid auction, the initial bids are not decisive; but like a sealed-bid auction, they are made simultaneously, albeit in two different stages. Recall in section 8.3 that we introduced a second stage in divide-the-dollar with two players, wherein information about the stage I bids made the strategy of choosing 50 by each player in stage II "dominance-inducible" under DD2.

To return to two-stage auctions, define a *sincere bid* as one in which a player bids his or her estimate of the true worth of an object, or his or her *valuation*, in stage I, making him or her indifferent between winning at that bid and losing the auction. In a two-stage auction, we will show that it is rational under plausible conditions for players to bid sincerely in stage I.

Define the players' valuations to be *private* – and the auction a *private-value*

[5] The informational advantages of conducting auctions in stages are discussed in Englebrecht-Wiggans (1988), but no specific procedure is recommended. Under a rather complicated set of rules, the Federal Communications Commission has recently conducted auctions for airwaves in stages (Passell, 1994; Andrews, 1995).

[6] In Brams and Taylor (1994d), we relax the assumption about valuation information to obtain Nash-equilibrium bidding strategies.

[7] In the latter case, if the bidders who tie at stage II also tied by making the same highest bid at stage I, the tie might be broken randomly. But the possibility of such a double tie seems remote unless players' valuations are not private (more on this possibility later). Note that if the purpose of the auction is to minimize the cost of a service to be performed, then "highest" would be replaced by "lowest" in rule 3. In this case, the players would be competing to make the lowest bid.

one – if these valuations are strictly their own and completely unaffected by their knowledge of the other players' valuations or bids. In particular, no player will alter his or her estimate of the value of an object in stage II once that player knows what the other players bid in stage I. In this situation, we will show in section 9.4 that, under plausible conditions, all players except the highest bidder will affirm their stage I bids in stage II; the highest bidder will usurp the second-highest bid, winning the auction at a price lower than his or her valuation and thereby ensuring himself or herself of a profit.

This outcome is identical to that in a *Vickrey (or second-price) auction*, which is a sealed-bid auction in which the player with highest bid wins but pays only the second-highest bid (Vickrey, 1961). Unlike a Vickrey auction, however, this outcome in a two-stage auction, which we call the *Vickrey outcome*, is not the product of dominant strategy choices by the players.[8] Rather, it is based on the successive elimination of dominated strategies, which is the same solution concept we used in our first revision (i.e., DD1) of divide-the-dollar (section 8.2).

While two-stage auctions implement the Vickrey outcome under rather general conditions, they do so in a way that obviates certain practical problems of Vickrey auctions that have made them rarely used (Smith, 1987; Rothkopf, Teisberg, and Kahn, 1990). Thus, by allowing the highest stage I bidder in a two-stage auction to learn the second-highest bid in stage II, he or she is able to choose whether to usurp it or not, whereas in a Vickrey auction the highest stage I bidder *must* pay this bid.

Two-stage auctions also provide a public means (to be described later) for the elimination of cheating or fraud in a Vickrey auction. Fraud can occur in a

[8] To demonstrate that sincere bids are dominant in a Vickrey auction, consider first the consequences of overbidding (i.e., above one's valuation). If one's overbid is the highest bid, and the second-highest bid is above one's valuation, then one wins but overpays; if the second highest bid is below one's valuation, then one could have won at the same price by sincere bidding, so sincere bidding is never worse and sometimes better than overbidding. On the other hand, underbidding may allow someone else to win when a sincere bid would have prevented this. Specifically, if the winner's bid is below one's valuation (and above one's underbid), then one loses out on what would have been a profitable sale. Hence, underbidding is sometimes worse than sincere bidding; it is never better, because when one does win with underbidding, one would have won at the same price with sincere bidding. Vickrey auctions, by separating what one bids from what one pays, afford one the opportunity, in effect, to be sincere without either overpaying or losing out to someone else on a profitable sale. English auctions, which are oral and ascending, lead to essentially the same result, because when no player makes a higher bid, one can obtain the item at a price only marginally higher than the second-highest bid. By contrast, Dutch auctions, which are oral and descending, and standard sealed-bid auctions are first-price auctions, because the winner must pay the highest price offered. Surprisingly, both types of second-price auctions (Vickrey and English), and both types of first-price auctions (Dutch and standard sealed-bid), all generate the same revenue for the bid taker in equilibrium (the so-called revenue-equivalence theorem), but this result breaks down if the key assumptions of the theorem (independently distributed private valuations and risk neutrality) are relaxed (Wilson, 1992; Kagel, 1995; Rothkopf, 1992).

Vickrey auction if bogus bids are surreptitiously introduced by confederates in order to increase the amount that the highest bidder must pay when he or she usurps the second-highest bid.

Even without bogus bids, winners in a Vickrey auction may end up paying almost what they bid, giving them only a minimal profit. By contrast, in a two-stage auction, unless *everyone else* bids only slightly less than the highest stage I bid, the highest bidder can effectively "bail out" by usurping a considerably lower bid. We demonstrate that this is not a rational strategy in a private-value auction, but in a *common-value auction* – in which everyone has the same valuation of an item (e.g., based on the price it commands in the marketplace), but about which there may be incomplete information – the ability of players to "correct" their bids in stage II attenuates the so-called winner's curse.

We believe two-stage auctions offer a persuasive and sensible alternative not only to Vickrey auctions but also to many other kinds of auctions that have been proposed or used (Smith, 1987; McAfee and McMillan, 1987; Milgrom, 1989). By making the stage I bids common knowledge in stage II, they alleviate the "unobservability-of-bidders'-valuations" problem alluded to earlier. In the process, they enable bidders to reduce uncertainty and make more informed and, therefore, intelligent choices in stage II, when their choices count. This is especially true in common-value auctions, in which players discover in stage II that they may have overbid (or underbid) in stage I, but first we analyze the private-value case.

9.3 Two-stage auctions: the private-value case

To begin the analysis, we make the following assumption:

> *Judicious-Bidding Assumption (JBA)*
>
> *In a two-stage auction, a player will never make a stage I bid, or choose a stage II option, such that, if that player wins the auction, he or she might suffer a loss.*[9]

An immediate implication of this assumption is that no player will ever overbid in stage I, because he or she might suffer a loss. A simple example illustrates this possibility. Assume that three players value an object at 1, 2, and 3, respectively. But they bid 4, 5, and 6 for it in order to try to ensure that their bids are the highest in stage I and, therefore, will be able to "trump" the choice of a competitor, should he or she choose the same option in stage II.

Once these bids are revealed in stage I, however, this common knowledge

[9] The use of the word "bid" in the stage I is clear; we use "option" rather than "strategy" in stage II in order to reserve the latter term for player choices in both stages of the auction. Thus, a strategy comprises both a bid and an option.

cannot help any of the three players, if he or she wins, avoid a loss. In particular, the players who bid 4 and 5, by affirming and usurping 4, respectively, can guarantee that the player who bid 6 wins but suffers a loss of 4 – 3 = 1 (when this player usurps the lowest bid of 4, which is the best he or she can do).

To avoid the possibility of overpaying, therefore, it seems reasonable to suppose that players will not make stage I bids greater than their valuations. Otherwise, they may lose; indeed, for the highest bidder, a loss may be unavoidable, as we just illustrated. JBA also precludes a player from usurping a bid in stage II greater than his or her valuation.

Our second assumption postulates that players want first to win at the minimum price, but then to be as competitive as possible in their bidding at this price. It requires the following definition: suppose that S and T are strategies for a player, Bob. Then we shall say that Bob *prefers* S to T if

- S weakly dominates T for him in the usual sense (section 8.2); or
- S and T are equivalent in the sense of yielding exactly the same outcomes, but S involves a higher stage I bid than does T, or the same stage I bid and a higher stage II bid.

Competitiveness Assumption (CA)

In a two-stage auction, players choose preferred strategies by successively eliminating nonpreferred strategies.

CA can be justified in the following way. Bob's primary goal is to win and pay as little as possible; all strategies that satisfy this goal are then distinguished by the secondary goal, which is to be competitive by bidding higher. This gives a lexicographic order, whereby the secondary goal is used to break "ties" between bids that satisfy the primary goal.[10]

Bob never suffers when he behaves according to CA. When all players, other things being equal, choose higher rather than lower bids (up to their valuations, because of JBA), the consequence is the iterative elimination of nonpreferred strategies, as we prove later.

Although CA seems unimpeachable as a prescription of rational behavior, it is rendered even more compelling if play is likely to be repeated. For example, we spoke in section 9.1 of a university in a town with several roofing contractors vying for roofing jobs at the university each year. Bidding strategies that stray very far from CA will certainly leave a contractor at risk of not even being invited to bid on future jobs.

Thus, even if a contractor knows that a competitor is sufficiently "hungry" to be unbeatable on a particular job, the contractor will want the university to regard him or her as one who comes forward with reasonable prices, even

[10] For further information on lexicographic orders, see Fishburn (1974a).

though he or she loses the bid on occasion. This is precisely the behavior CA prescribes.

Although the rationale of CA rests on the assumption of repeated play, we model two-stage auctions as a single-play game. In effect, CA provides a rationale for competitive bidding in such a game.

Given JBA and CA and two minor technical conditions, a two-stage auction turns out to be equivalent to a Vickrey auction in the sense that both elicit sincere bids and make the highest bidder the winner, paying the second-highest bid. We shall argue later that the *way* in which a two-stage auction implements the Vickrey outcome confers on it substantial practical advantages, but first we state Theorem 9.1, whose proof we give in the appendix to this chapter.

Theorem 9.1 Consider a two-stage auction with n players. Assume the following are true and common knowledge (section 4.2):

1 *Different players never have equal valuations and never make equal bids in stage I.*
2 *All reservation prices are in the interval [0, m], and the value of m is common knowledge.*[11]
3 *JBA, and that it is common knowledge.*
4 *CA, and that it is common knowledge.*

Then iterative elimination of weakly dominated strategies leads all players to bid their valuations in stage I. In stage II, the highest bidder will usurp the second-highest bid and thereby win the auction, thus implementing the Vickrey outcome.

CA is not the only assumption, in combination with the other three in the statement of Theorem 9.1, that can be used to prove the rationality of the Vickrey outcome in a two-stage auction. As an alternative assumption, consider the following:

No-Regret Assumption (NRA)

In a two-stage auction, a player will never make a stage I bid, or choose a stage II option, such that, if he or she should turn out to be the player with the highest valuation, he or she might lose the auction.

Consider NRA in conjunction with JBA. By JBA, Bob will never bid more than his valuation in stage I. By NRA, he will never bid less than his valuation; otherwise, if he is the highest bidder, his stage I bid might not be the highest,

[11] If the lowest bid is the winning bid (e.g., in a contracting job), then "0" would play the role of "*m*."

and he will definitely lose if the *highest bidder affirms*. Consequently, Bob will bid his valuation in stage I.

Notice that this argument for sincere bidding at stage I postulates the *possibility* that the highest bidder will affirm in stage II. While this possibility may not be rational (e.g., if this highest bidder bids his or her valuation), it induces everyone to be sincere in stage I. But if everyone is sincere, and everyone knows everyone is sincere (which they do because JBA and NRA are common knowledge), then the highest bidder will know that he is and usurp the next-highest bid.

With NRA instead of CA, then, we do not need to rely on any inductive arguments, based on iterative weak dominance, to prove the rationality of the Vickrey outcome. It instead follows by contradiction: assuming that the highest bidder might affirm, we show that this will never be the case because:

- all players will be sincere in stage I, given JBA and NRA;
- in stage II, the highest bidder (assume this person is Bob) will therefore know that he is the highest bidder;
- knowing that he is, he can usurp the next-highest bid, assured that he will not be usurped.

What are the relative merits of substituting NRA for CA in Theorem 9.1 to establish the rationality of the Vickrey outcome? On the plus side, NRA does not require the use of a long chain of reasoning – the five claims in the proof of Theorem 9.1 in the appendix – to yield the Vickrey outcome. Rather, the rationality of this outcome follows immediately from NRA and JBA.

On the minus side, NRA does not really justify the rationality of the Vickrey outcome. Rather, it simply asserts that players will not underbid, but only because they fear they might lose if they should turn out to be the highest bidder and thereby experience regret.

But consider a player (assume this person is Carol) who is almost certain she is not the highest bidder (assume this person is Bob) but thinks she is probably the player with the second-highest valuation. Moreover, suppose Carol thinks that the highest bidder will substantially underbid in stage I. Then if Carol only slightly underbids in stage I *and* turns out to be the highest bidder, she should affirm in stage II (because the highest bidder will definitely usurp her stage II bid), ensuring herself of a positive, if small, profit.

Because this rational scenario for Carol is inconsistent with her bidding sincerely in stage I (and usurping in stage II), it casts doubt on the reasonableness of NRA. CA, by contrast, provides rational reasons for players to be sincere in stage I and, consequently, for Bob, necessarily the highest bidder, to usurp the second-highest bid in stage II.

We conclude that CA offers a more persuasive justification of the Vickrey outcome from "first principles": successively eliminating weakly dominated strategies enables players to maximize their payoffs should they win. While

NRA ignores profit maximization – indeed, the player with the second-highest valuation can suffer if the player with the highest valuation underbids and the second-highest valuation player does not – NRA is nevertheless quite commonsensical. It, along with JBA, provides players who want to avoid possible regret a transparent reason to bid sincerely and then choose the Vickrey outcome.

The transparency of NRA must be weighed against the more full-fledged justification, from first principles, of CA. We do not takes sides on which assumption is "better" but simply point out that there are different routes that lead to the consequence of Theorem 9.1.[12] Moreover, both lead to an efficient outcome: the bidder with the highest valuation wins the auction.

To summarize, we analyzed private-value two-stage auctions – in which each player has his or her own valuation of the object being auctioned off, based on private information – that does not depend on the values of other players: a player's valuation would not change if these other valuations became common knowledge. We argued that, for two different sets of assumptions, the players would make sincere bids in stage I, and the highest bidder would usurp the second-highest bid in stage II and win, with the formal argument given in the appendix to this chapter.

This result duplicates the outcome of a Vickrey auction, which is a sealed-bid auction in which the highest bidder wins but pays only the second-highest bid. But if two-stage auctions simply implement the Vickrey outcome, it is legitimate to ask what advantages they offer that Vickrey auctions do not.

We think there are several. First, they provide an easy way – by means of a simple checking procedure (to be described next) – to prevent the introduction of bogus bids by shills or confederates that can raise the amount that the highest bidder must pay.

Under this checking procedure, one would ask each player – whose legitimacy as a bidder is known or can be verified – at the end of stage I whether his or her bid is one of those made public. If each bidder answers affirmatively (assume that all the bids are different and equal to the number of players), one can be sure that no fraudulent bids have been inserted to bias the

[12] Another path is to assume a different tie-breaking rule in stage II: in the event of a tie for the highest bid, the winner is the player who has the closest but strictly higher bid in stage I. Then claim 1 of Theorem 9.1 would work by a different induction argument: the highest stage I bidder will not usurp the lowest bid because he or she will definitely lose, and hence no player will usurp this bid; proceed inductively (highest bidder will not usurp the next-to-lowest bid, etc.) until the next-highest bid is reached, which will be usurped by the highest bidder. Note that if $n = 2$, this tie-breaking rule is the same as that assumed in the text (i.e., the highest stage I bidder breaks a tie in his or her favor). In Brams and Taylor (1994d), we show that the strategies of the players given by Theorem 9.1 constitute a symmetric Nash equilibrium under rather weak conditions, which is a unique equilibrium under slightly stronger conditions.

outcome in favor of obtaining a higher price for the bid taker. This procedure would address one of the principal objections – fear of cheating – that Rothkopf, Teisberg, and Kahn (1990) and Rothkopf and Harstad (1995) claim bedevils Vickrey auctions and has worked against their adoption, though collusion among a ring of bidders may still be possible (Mailath and Zemsky, 1991).

To be sure, such a checking procedure could be used in a Vickrey auction. But it would require the public revelation of all bids, which is automatic in two-stage auctions but would have to be added as a new feature of Vickrey auctions.

A far more important advantage of two-stage auctions is that they allow bidders to revise their bids by usurping others' bids once the latter become common knowledge. True, we showed that all bidders except the highest will be motivated to affirm their stage I bids in private-value auctions. But in common-value auctions, in which the option one chooses in stage II may depend on the other stage I bids (because one's valuation is not private – it depends on information about others' valuations or bids), one can conceive of situations in which a player, after observing the bids in stage I, decides that he or she wants to opt out, or at least to revise his or her bid.

Before considering how this bidder might do so in two-stage auctions, we return to the subject of one-stage auctions. We analyze the well-known winner's curse in these auctions and later show how two-stage auctions would mitigate this problem and also speed up the search for the so-called common value of an item.

9.4 One-stage common-value auctions: the winner's curse

Assume that there is a single common value that an object is worth (e.g., its price in a market), but the players can only estimate its value. Such auctions are probably the norm; even art has its common-value aspects (McAfee and McMillan, 1987, p. 726).

We begin by modeling how players estimate common value in *one-stage* sealed-bid auctions. Assume there are n players 1, 2, . . . , n in a sealed-bid auction. Their *initial estimates* of the common value of the object being auctioned off are $v_1(1), \ldots, v_1(n)$, where we use the subscript "1" to indicate that these estimates are only preliminary and will be revised later. If the auction is of an oil tract, for example, these estimates might be the best judgments of each company's geologists, based on their private information, of the tract's worth.

We assume that each company associates with its judgment a fraction α, where $0 \leq \alpha \leq 1$, which indicates its level of confidence in its geologists' estimate. The complementary fraction, $1 - \alpha$, indicates its confidence in all the other companies' estimates, which are averaged and considered as a collectivity. For simplicity, we assume that all players have the same α, which may or may not be realistic.[13]

If $\alpha = 1$, the auction is one of independent private values. In this case, of course, the judgments of the other players do not affect a player's estimate.

It might be thought that these judgments also do not matter in a one-stage auction, because a player does not learn about the other players' bids – much less their geologists' estimates – until the auction is over, when it is too late for this player to revise his or her estimate. But, as we shall next show, simply an *appreciation* of the winner's curse can cause a player to revise his or her initial estimate and, as a consequence, alter his or her bidding strategy.

Suppose that each player bids this initial estimate, and player 1 wins with his or her bid of $v_1(1)$. Then player 1 knows that the other $n-1$ bids – $v_1(2), \ldots, v_1(n)$ – are all less than $v_1(1)$. This knowledge enables player 1 to make a revised calculation of the value of the tract, taking into account not only what his or her own geologists say but also what the geologists of the other $n-1$ players must have said. To wit, player 1's revised estimate, denoted $v_2(1)$, is

$$v_2(1) = \alpha v_1(1) + (1-\alpha)\{[v_1(2) + \cdots + v_1(n)]/(n-1)\}. \tag{9.1}$$

To be sure, without making additional assumptions, player 1 does not know the values of the $n-1$ terms in brackets on the right-hand side of (9.1). Nevertheless, assuming $\alpha < 1$, he or she can surmise that $v_2(1) < v_1(1)$. That is, if player 1 bids his or her initial estimate of $v_1(1)$ and wins, player 1 is paying more than the tract is worth, at least according to the collective wisdom of all the geologists involved. This is the *winner's curse* in our model: the winner is cursed by overpaying if he or she bids an initial estimate.

To avert the winner's curse, suppose that player 1 revises his or her initial estimate, based on the following thought experiment: player 1 assumes, provisionally, that:

(i) the other players bid sincerely [i.e., bid their initial estimates of $v_1(j)$ for $j = 2, \ldots, n$]; and, for concreteness, that
(ii) these bids are uniformly distributed below his or her bid, should this bid be sincere and winning.

Then player 1 can determine that the average of the other bids, given in braces in (9.1), is $v_1(1)/2$.[14]

[13] We will consider the reasonableness of this assumption later in two-stage auctions, where we shall suggest different measures for taking account of the estimates of others.

[14] More generally, this will be the average if the bids are *symmetrically* distributed below $v_1(1)$. We make the more restrictive assumption of uniformity here because we use it later in the calculation of a revised estimate; this assumption is also made in a discounting model described in Rasmusen (1994, p. 296). Neither the assumption of asymmetry nor of uniformity is realistic, however, if bids tend to cluster (e.g., just below player 1's winning bid), which is a phenomenon we discuss in section 9.5 in connection with common-value auctions.

Substituting this average into (9.1), player 1 will revise his or her initial estimate as follows:

$$v_2(1) = v_1(1) + (1 - \alpha)[v_1(1)/2].$$

Because this calculation pertains to every player, we can delete the "(1)," obtaining:

$$v_2 = \alpha v_1 + (1 - \alpha)(v_1/2) = (v_1/2)(\alpha + 1). \tag{9.2}$$

This revised estimate, it would seem, will enable the players to avert the winner's curse – at least to the extent that any probabilistic considerations can provide help. Notice, in particular, the behavior of v_2 at the extremes: when $\alpha = 0$ (i.e., only the *other* players' bids matter), players will bid exactly one-half their sincere estimates; when $\alpha = 1$ (i.e., only *one's own* estimate matters), players will bid exactly this estimate. Because the estimates of players generally depend on both their own estimates and those of other players, this mixture will lead them to bid below their own initial estimates, according to (9.2).

But this finding depends on assumption (i) – that "the other players bid sincerely" – which is inconsistent with our assumption that player 1 will revise (i.e., lower) his or her v_1 estimate to v_2. One cannot have it both ways: either all players make a revised estimate v_2 (as assumed of player 1), or no players do.

If no players do, we are back to our original estimates ($\alpha = 1$), and the winner's curse remains. If all players do, we must take this fact into account by amending assumption (i).

We do this by postulating that *all* players go through the same thought experiment as player 1. Thereby, we do not single out player 1 as special but instead assume that the $n - 1$ other players also revise their v_1 estimates to v_2's, and player 1 knows this.

This knowledge on the part of player 1 leads him or her to make a still higher-order estimate, whereby player 1 revises $v_2(1)$. We call this revision $v_3(1)$; it assumes that every other player has made second-order estimates $v_2(j)$ for $j = 2, \ldots n$, which are then factored into player 1's third-order estimate of $v_3(1)$.

To calculate this estimate, we assume that player 1 gives a weight of α to v_1, not v_2, because v_1 is *his or her* geologist's estimate of true worth. But now player 1 must take into account that, if he or she wins, this is because his or her bid exceeds the v_2's of the $n - 1$ other players, not their v_1's.

If all players bid their v_2's, and player 1 wins, the bids of the other players are no longer uniformly distributed over $[0, v_1(1)]$ but over the smaller interval $[0, v_2(1)]$. From (9.2) and the fact that this distribution is uniform, we can normalize the smaller interval by multiplying by the factor $[2/(\alpha + 1)]$. Player 1's third-order estimate, based on his or her own first-order estimate and the second-order estimates of the other players, then becomes:

$$v_3(1) = \alpha v_1(1) + (1 - \alpha)[2/(\alpha + 1)] \text{ [expected value of other players' bids, given player 1 wins with bid of } v_2(1)]$$
$$= \alpha v_1(1) + (1 - \alpha)[2/(\alpha + 1)][(1/2)v_2(1)]$$
$$= \alpha v_1(1) + (1 - \alpha)[1/(\alpha + 1)]\{[v_1(1)/2](\alpha + 1)\}$$
$$= \alpha v_1(1) + (1 - \alpha)[v_1(1)/2] = v_2(1)$$

from (9.2). Lo and behold, the third-order estimate leaves the second-order estimate unchanged!

Thus, the whole process stabilizes once the players reach the third order of sophistication. Consequently, after the second order of sophistication, the players can rest assured that their downward revisions, as given by (9.2), probabilistically avert the winner's curse, and no further adjustments are necessary.

A good auction procedure should enable players to exploit as much information as possible before settling on a bidding strategy. Sealed-bid auctions, because they occur in only one stage, are flawed in this regard. By permitting only *a priori* calculations of the kind we have modeled, they deny players feedback on other players' valuations, on the basis of which they can revise.

Sophisticated players today probably have at least an intuitive understanding of the winner's curse (McAfee and McMillan, 1987, p. 730). Whether they depreciate their initial estimates in the manner we have described or discount them in other ways, there is good reason to believe that their bids are not sincere.

While Vickrey auctions offer one solution to this problem, in the common-value case they offer no help to players in acquiring additional information on which to base a common-value estimate. Although they may relieve players of the need to discount their initial estimates, they do not enable them to incorporate into these estimates information about the estimates of other players.

Is there a better way to avert the winner's curse? We think, by providing players with information on other players' initial estimates, that there is.

9.5 Two-stage auctions: the common-value case

Two-stage auctions differ qualitatively from one-stage auctions in having a stage I that "does not count." Of course, this is not literally true, because the stage I bids provide the menu from which players select their bids in stage II as well as determine the winner if there is a tie in stage II. And what gets on the menu and who ties certainly affect what gets chosen in the end.

Nevertheless, a player never has to pay what he or she bids at stage I, so players make no binding commitments at this stage. Thus, if a player is the highest bidder at stage I, this player can usurp any lower bid that he or she thinks might be a more reasonable estimate of the common value; if this player wins, he or she will only have to pay this bid. If a player is the lowest bidder, this player is even less committed, because affirming his or her own bid ensures that he or she loses, no matter what the other players do.

The evaporation of commitment in stage I would seem to afford every player the opportunity to be sincere at this stage – duplicating the results for private-value two-stage auctions (section 9.3) – rather than worry about the winner's curse and try to adjust for it in his or her initial bidding. After all, what is there to lose if a player (assume this person is Bob) can always "bail out" in stage II with a lower bid, even if he is the highest bidder at stage I, or move up and usurp a higher bid?

To investigate the robustness of sincerity in stage I, we need to make some assumptions about the common value of an object. Unlike one-stage auctions, we assume that Bob's initial estimate will not, in general, be determinative. On the contrary, we assume that the other bids that Bob observes in stage II not only matter but also play a critical role in his choice of a stage II bid. After all, the estimate of Bob's geologists is only one of several; some of these might be just as good as, if not better than, his own estimate.

To give the estimation problem greater structure, assume that players believe that the best estimate of the common value of an object is the median bid in stage I – that is, this is most likely to be its true value. Therefore, the most competitive yet profitable bid they can make in stage II is the next-lower bid to the median.

The choice of the median as the best estimate in stage II is arbitrary, even when all the other players are sincere in stage I. Moreover, the choice of the next-lower bid as an appropriate amount to bid in stage II seems questionable in certain situations. For example, consider the two cases, Cl and C2, in both of which 95 is the median bid (underscored) – half the bids lie below this bid and half above:

C1: stage I bids are 99, 98, 97, 96, <u>95</u>, 4, 3, 2, 1

C2: stage I bids are 99, 98, 97, 96, <u>95</u>, 94, 3, 2, 1

In C1, the 99-player would win at 4, whereas in C2 this player would win at 94 if, as we provisionally assume, all players usurp (or affirm) the next-lower bid to the median.

Notice that these vastly different bids stem from only one different bid in the two cases. Intuition suggests that with this much on the line, players might invoke other estimation procedures, based on standards other than the median.

For example, what if the mean, which is 55 in Cl and 65 in C2, were the best estimate? As before, the 99-player would usurp 4 as his or her first profitable bid in C1; in C2, however, the 99-player would usurp 3, not 94. Even if one believed in the median as the appropriate standard, bidding 94 in C2 seems an extremely tough call because of its enormous significance for profits.

It is not surprising that the median and the mean may give different results. Still other standards may be appropriate; for example, elsewhere we propose a measure called the "weighted point estimate," which has properties of both the median and the mean (Brams and Taylor, 1994e). But here we shall investigate implications of using the median of the stage I bids as the best estimate of the common value of an object:

Theorem 9.2 Consider a two-stage auction in which all players use the following strategy:

S: be sincere in stage I and usurp the highest bid that is below the median bid in stage II.

Assume that the median sincere bid is the best estimate of common value. Then the players who can benefit by a unilateral defection from S in stage I are precisely those whose sincere bids are strictly between the highest stage I bid and the winning stage II bid (which is at least the median, but not the highest, stage I bid). Moreover, all these players can equally benefit either by leapfrogging the lowest bid (i.e., bidding below it), or leapfrogging the highest bid (i.e., bidding above it), made by others in stage I.

Proof. Suppose first that there are an odd number of bidders. Let m be the median bid, and let b be the bid immediately below m. We consider the effects of a unilateral defection by a player P who bids x, which we let range over all the bids:

Case 1. P leapfrogs the lowest bid in stage I.[15]

1.1 $x < b$. In this case, the highest bidder still wins at b.

1.2 $x = b$. If b is the lowest bid, this case is vacuous. If not, then the median is unchanged, but the bid b' immediately below b becomes the bid usurped by every player using S. Thus, P would have to usurp m in order to win (because there are no bids between b' and m once P has defected). But this win at m is the same as losing for P.

1.3 $m \leq x$ and x not the highest bid. In this case, b becomes the new median, so every player who uses S will usurp a bid below b. Thus, P can usurp b and win at a profit.

1.4 x is the highest bid. If P proceeds as in case 1.3, he or she wins at b exactly as P did with the highest bid.

[15] We do not assume that P knows the lowest bid but, rather, if he or she made this choice (unknowingly), there would be the four possible consequences that we list next.

Case 2. P leapfrogs the highest bid in stage I.

2.1 $x \leq b$. In this case, the bid above m becomes the new median, so every player will usurp m. P can win by usurping m, but this is the same as losing the auction.

2.2 $x = m$. In this case, the bid above m again becomes the new median, but now every player will usurp b. Thus, P (or the new highest bidder) can win by also usurping b.

2.3 $m < x$. If x is the highest bid, this case is vacuous. Otherwise, neither the median nor the price changes, but P becomes the new winner at b.

A similar argument applies if the number of bidders is even, with the median the average of the middle two bids. Q.E.D.

Corollary 9.1 In a two-stage auction with two players, S is a Nash equilibrium when the median (or the mean) is used as the best estimate of common value.

Thus, when we limit two-stage auctions to two players, we get stability in the sense of Nash (section 4.3). When there are more than two players, the fact that a player whose bid is the median or above (but not the highest) can win – by departing either up or down – suggests that there is no fundamental bias favoring upward or downward departures. To illustrate Theorem 9.2, we consider a three-person example.

Let the set of stage I sincere bids be {l, 3, 11}, so $m = 3$ is the median bid. Call the median bidder M. In stage I, M can insincerely leapfrog up to, say, 12; in stage II, M can then usurp the bid of l, the bid immediately below the new median, $m' = 11$, and win. Alternatively, M can leapfrog down to 0, in which case $m' = 1$, and the 11-player, following S, will usurp 0 in stage II. But at this point M can usurp l and win.

Unlike M, the 11-player has no reason to depart from S when everybody else chooses this strategy, because he or she can win by usurping the bid of l. Neither does the 1-player have reason to depart, because no insincere bid in stage I – either higher or lower than 1 – can make him or her the winning bidder in stage II at a lower value than 3, which is equivalent to losing the auction.

A comparison of the median with the mean as best estimator is instructive. In the two-person case, they give the same result: by Corollary 9.1, strategy S is a Nash equilibrium.

Differences crop up between the median and the mean as best estimators when there are more than two players. Consider a strategy analogous to S when the mean is the best estimator:

S': be sincere in stage I and usurp the highest bid that is below the mean bid in stage II.

Returning to our previous example, the mean of {l, 3, 11} is 5. Hence, following S', the 11-player will usurp 3 (not 1, as in the case of the median) and win with a profit (5 – 3 = 2).

As in the case of the median, the three-player, by insincerely bidding 12 in stage I, can benefit. The new mean will be 8, so following S' the insincere three-player (now the 12-player) will usurp the bid of 1 and win with a profit (5 – 1 = 4). However, if 0 is a lower bound on bidding (i.e., no negative bids are allowed), the three-player cannot win by underbidding. The reason is that by bidding 0 in stage I, the three-player lowers the mean only to 4. Consequently, the 11-player will still usurp 3 and win, precluding a win for the insincere three-player (now 0-player) at 1 (or 3).

But this apparent bias in favor of insincerely bidding up is simply an artifact of the particular numerical bids. If they were each increased by, say, 10, making them {11, 13, 21}, then if the 13-player insincerely bid 0 in stage I, his or her bid would lower the mean from 15 to $10^2/_3$. Then the 21-player, following S', would usurp 0 rather than 11, enabling the 13-player (now the 0-player) to usurp 11 and win with a profit (13 – 11 = 2).

The lesson in these cases is that insincere bidding in stage I may be profitable for the middle bidder, whether the best estimate is the median or the mean. By leapfrogging either the highest or the lowest bids in stage I, the middle bidder can induce the other players, following S or S', to behave in a way that he or she can exploit. Indeed, our last example with the mean – and our earlier example with the median – demonstrated that the insincere middle bidder may even succeed in winning at the lowest sincere bid, giving him or her maximal profit.[16]

Whether the best estimator is the median or the mean, the lack of a Nash equilibrium for $n > 2$ would seem to jeopardize sincere bidding in stage I. As we have seen, however, it is by no means evident whether players, especially when information is incomplete (as we assume in stage I), can do better by underbidding or overbidding.

So far we have shown that, using the median or the mean as a best estimator, sincerity in stage I is not in general optimal in common-value auctions. Some players will have an incentive to dissemble, but it is not apparent how, in a game of incomplete information, they will be able to ascertain whether they are in a position to benefit – and how to do so – when they make their stage I bids. By Ockham's razor, therefore, they may as well be sincere.[17]

[16] If the mean is the best estimate, the *lowest* sincere bidder – not just the middle bidder – can also drag the mean below his or her own sincere bid by substantially underbidding in stage I. This bidder can then win at the lowest sincere bid in stage II (i.e., his or her own) when the higher bidder usurps his or her lower insincere bid.

[17] This may seem a strange argument for sincerity, but it is one that has been persuasively made in the social-choice literature. That is, a voting system will induce sincerity if, though manipulable (as all voting systems are), voters have neither the incentive (e.g., because the

Even after players acquire information about the other players' stage I bids, it is also not clear that they will all rush to affirm or usurp the same bid just below the median or mean. If there are ten bidders, for example, and five bids are tightly clustered around the median or mean, which should a player choose? True, the highest stage I bidder has an advantage if there is a tie, but knowing this, the other players may have an incentive to choose one of the higher bids in this cluster. But then the highest bidder can anticipate this action and may also aim higher.

In light of this rather confusing picture at stage II, consider the strategic problem at stage I. Perhaps the best recourse of a player is to be sincere – bid the initial estimate of its geologists in the case of the oil tracts. But as the bids for offshore oil tracts in table 9.1 demonstrate, there may be little agreement on what the true values of these tracts are, assuming these are sincere bids. If they are not – perhaps because they have been discounted according to a model like that presented in section 9.4 – one would still not expect the estimates to be so wildly different unless the science of estimating oil potential is truly primitive.

This may have been so, at least as of the late 1960s. Moreover, these bids, which differ by a factor of as much as one hundred, were made before the winner's curse was first described by three Atlantic Richfield engineers, Capen, Clapp, and Campbell (1971).[18] In the auction of U.S. Treasury debt notes, the story is far different, where a spread of six basis points (hundredths of a percentage point) is considered very large, though it is only one-hundredth of the value of the notes being auctioned off (Gilpin, 1991).

The new information that the revelation of stage I bids provides, especially if the bids are sincere, makes all players better informed about each others' initial estimates. In this sense, it is a public good, enabling the players better to avert the winner's curse.

Although a Vickrey auction would partially alleviate this problem in a one-stage sealed-bid auction, if the two highest bids (among, say, ten) are very close, the highest bidder may still end up overpaying at the second-highest bid. Why? Because in a common-value auction, such a bid might be viewed, in light of the eight lower bids, as simply "too high."

By contrast, in a two-stage auction, the highest bidder might well usurp the

calculations are too complicated to make) nor the ability (e.g., because of incomplete information) to effect outcomes that could benefit them. The Hare system of single transferable vote (STV) (section 10.2) is a good example of a voting system that is theoretically manipulable but, in practice, would be difficult to exploit, as we argued was also true of AW (section 4.3).

[18] It is perhaps not surprising that financial returns from these leases were quite dismal, strongly pointing to the existence of a winner's curse (Thaler, 1988, 1992); how one large oil company (Gulf Oil Corporation) attempted to deal with this problem in the early 1980s is discussed in candid detail in Keefer, Smith, and Back (1991). Laboratory experiments on winner's curse, as well as other aspects of auctions on which experiments have been conducted, are reviewed in Kagel (1994).

Table 9.1. *Bids (in millions of dollars) by "serious competitors" in oil auctions*

Offshore Louisiana, 1967 (Tract SS 207)	Santa Barbara Channel, 1968 (Tract 375)	Offshore Texas, 1968 (Tract 506)	Alaska North Slope, 1969 (Tract 253)
32.5	43.5	43.5	10.5
17.7	32.1	15.5	5.2
11.1	18.1	11.6	2.1
7.1	10.2	8.5	1.4
5.6	6.3	8.1	0.5
4.1		5.6	0.4
3.3		4.7	
		2.8	
		2.6	
		0.7	
		0.7	
		0.4	

Source: Capen, Clapp, and Campbell (1971), p. 642, Table I.

third-highest, fourth-highest, or fifth-highest bid. But whatever estimate of common value this bidder makes (median, mean, or something else), he or she – as well as everybody else – can at least make his or her own choice in a two-stage auction.[19]

Finally, it is worth noting that two-stage auctions allow for *split awards*, in which a divisible good is apportioned among two or more bidders (Anton and Yao, 1989, 1992; Klotz and Chatterjee, 1995). In fact, we believe that split awards might be better wedded to two-stage auctions than one-stage auctions. While splitting an award in a predetermined ratio between the two top bidders reduces the risk for each in a one-stage auction, it raises the problem of price discrimination if one player receives a portion of the award at a lower price than the other.

By contrast, with two-stage auctions, if there is a tie in stage II, one can make an award to the tied players at the same price. In addition, instead of fixing the proportions exogenously, one can determine them endogenously, based on

[19] Another way to avoid the winner's curse is to allow players to buy insurance that enables them, if they turn out to be the highest bidder, to withdraw their bids and lose the auction. More directly, one could charge the highest bidder a penalty for withdrawing his or her bid. However, giving players this option makes them more aggressive in their bidding (Harstad and Rothkopf, 1995) – as does making the winner's payment a weighted sum of the two highest bids (Riley, 1988) – which renders estimates of common value more problematic. Two-stage auctions, in our opinion, do a better job of providing a bail-out option while fostering sincere bids (more on this point shortly).

the stage I bids. Presumably, the highest stage I bidder would win a greater portion of the award than lower stage I bidders who tie in stage II (we leave open the function of their stage I bids that might be used, which would affect stage I bids). Two-stage auctions, then, would become a special case of two-stage split auctions, in which the split is 100 percent to the highest bidder.

We conclude that the instability of common-value two-stage auctions is not necessarily damning, because it is not biased in either direction. Although it would appear that players would always want to bid higher than their sincere estimates to be able to break ties in their favor in stage II, we showed that players, on occasion, may do better by bidding lower.

The lack of a systematic bias, inducing the players either to overbid or to underbid in stage I, gives them good reason to be sincere in this stage instead of being "too clever by half." Then, after the incomplete information of stage I becomes more complete in stage II, the players are in a better position to home in on the common value.

9.6 Conclusions

We began our analysis of two-stage auctions by saying that they were a way of implementing the Vickrey outcome. Given the Judicious-Bidding Assumption (JBA) and the Competitiveness Assumption (CA), we showed that players in the private-value case would not only bid sincerely in stage I but also that the highest bidder (with the highest valuation) would usurp the second-highest bid in stage II.

Substituting the No-Regret Assumption (NRA) for CA makes this result more transparent. But NRA, in precluding underbidding in stage I, is not rooted to fundamental rationality calculations (e.g., based on the desire of players to maximize their payoffs). Rather, it asserts that players will eschew the possibility of any regret that might occur if, by underbidding, they lose. But blindly following NRA could cost players profits in certain scenarios, whereas CA rules out such scenarios when the players themselves successively eliminate dominated strategies.

While two-stage auctions, like Vickrey auctions, induce players initially to bid sincerely in the private-value case, they offer, in our view, important practical advantages over Vickrey auctions. In particular, they allow players to revise their bids in stage II in light of knowledge gained when the other bids are revealed in stage I, which is especially important in the common-value case.

In one-stage auctions, by comparison, players must try to take account of the winner's curse in which a player, because he or she wins, overpays. To avert this curse – that is, the "bad news" about winning – we argued in section 9.4 that sophisticated players will generally depreciate their sincere estimates of the worth of the item being auctioned off by factoring into their calculations how other players revise their own initial estimates, resulting in a lowering of all bids.

To be sure, even in two-stage auctions, the information players learn at the conclusion of stage I may still leave them in a quandary if the auction is a common-value one. In such auctions, each player's valuation depends on the other players' valuations, as manifested in how they bid in stage I.

In section 9.5, we showed that sincere stage I bids are not generally a Nash equilibrium. Which players have an incentive to deviate, however, depends on whether players use the median, mean, or some other measure of first-stage bids to estimate the common value. Thus, if the median is used, then the only players who will have an incentive to be insincere are those whose bids are at or above the median but not the top bid. On the other hand, if the mean is considered to be the best estimate of common value, other players (possibly all except the top bidder) may have an incentive to defect from their sincere bids.

We do not believe that dilemmas of this kind necessarily undermine two-stage common-value auctions. Rather, they mirror the inherent difficulties of bidding in such auctions, which is clearly a risky business.[20]

This risk may actually contribute to sincerity. Because players cannot be sure whether they are in a position to do better by bidding higher or lower than their sincere initial estimates, they may be sincere by reason of default. At worst, departures from sincerity are unlikely to be systematically biased (e.g., all high or all low).

Sincerity in stage I bidding is reinforced by the fact that, unlike a Vickrey auction, the highest bidder need not pay the next-highest bid but can, instead, opt out or at least drop considerably lower in stage II. Or a low bidder can usurp a higher bid. Giving players such freedom, we believe, is not only sensible but also encourages sincerity in stage I.

But it is in stage II, after presumably sincere stage I bids are revealed, that the advantages of two-stage auctions are most striking. They not only give players an opportunity to bail out if they overbid, or usurp a higher bid if they underbid, in common-value auctions, but, more important, they give them a better opportunity to approximate the common value by seeing all the stage I bids at once.[21]

We believe that two-stage auctions, whether split among more than one high bidder or not, offer a promising alternative for players to approach, through their bidding, the common value of objects. We have not resolved all

[20] The fact that oil drilling at offshore tracts has turned up mostly dry wells indicates that oil companies do indeed take recognizable risks (Thaler, 1988, 1992).

[21] The oral, ascending bids of English auctions also give players the opportunity to revise their bids, but not on the basis of seeing – at any one time before the end – all the bids together on the table.

ambiguities at either stage I (will players be sincere?) or stage II (how will they estimate the common value?), which laboratory experiments with such auctions might help to clarify, facilitating comparisons with other auction procedures (Smith, 1987; Kagel, 1995). If the apparent advantages of two-stage auctions are borne out in the laboratory, we think they should be tried out in natural settings.

To gain such adoptions, however, a crucial question must first be answered: Will the bid taker benefit from a two-stage auction? If not, such auctions are not likely to be held. Surprisingly, both the bid taker and the bidders may benefit if the stage I bids are sincere:

- The bidders benefit if the winning stage II bid approximates the common value, less some reasonable profit. Then the procedure will be viewed as fair, which will help to legitimate the "rules of the game" and, at the same time, reward the winner with an appropriate profit;
- The bid taker benefits, at least over the long run, because the bidders neither (1) overbid, and then, because of their losses, become too conservative; or (2) underbid, and then, because of their gains, become too aggressive. Rather, a "steady state" is reached sooner than in a one-stage auction, giving the bid taker a greater expectation of consistently receiving a fair price – based on the common value – which he or she desires over "maximizing . . . commission revenue from a single auction" (Rothkopf and Harstad, 1994, p. 365).

True, what the bidders lose the bid taker gains – and vice versa – because of the zero-sum character of auctions. On the other hand, all parties gain, in a sense, if the prices of the goods being auctioned off reflect their common value. Because no party gets "taken for a ride" when this is the case, they are better able over time to build up stable expectations about the procedure and its presumptive fairness.

This presumption will be more palpable to bidders if there are different winners of different items in the same auction, or different winners of items that come up for bid over time in different auctions. Suppose, for example, that there are five serious building contractors in a city, who bid for different jobs at different times. If they have varying amounts of work at any time, the bids of those who have some slack, and can therefore better afford to take on new jobs, are more likely to be competitive than those who are already overworked.

In this situation, the winners will indeed vary as a function of the workload that the different contractors have. Over time, the jobs will, we believe, be distributed fairly among the different contractors, which in principle will enable everybody to share profitably in the building-contract business.[22]

[22] The fact that two-stage auctions would provide the players, via the stage I bids, with better information about each other's positions would, we believe, contribute to this fair division while

Moreover, if indeed the bidder with the highest valuation (e.g., the hungry contractor, or the collector enchanted by a particular painting) wins the auction, then other players should not envy him or her.

We turn to elections in chapter 10, where we will focus on those in which there are multiple winners, such as to a council or legislature.[23] The problem here is how to aggregate preferences across voters to determine a set of winners, which does not arise in auctions.

If there is an aggregation problem in two-stage auctions, it is how to determine the common value from the stage I bids. This problem is somewhat akin to the problem that a candidate faces in judging who, say, the median voter is to whom he or she should try to appeal in a campaign.

In multiple-winner elections, the natural fairness question is how best to ensure the proportional representation of different parties or other groups in a council or legislature. After reviewing extant systems, we will describe in some detail a new election system, called "constrained approval voting," that ensures the representation of two different kinds of characteristics of the electorate and also takes into account the popularity of different candidates.

Appendix

Theorem 9.1 Consider a two-stage auction with n players. Assume the following are true and common knowledge:

1 *Different players never have equal valuations and never make equal bids in stage I.*
2 *All reservation prices are in the interval [0, m], and the value of m is common knowledge.*
3 *JBA, and that it is common knowledge.*
4 *CA, and that it is common knowledge.*

Then iterative elimination of weakly dominated strategies leads all players to bid their valuations in stage I. In stage II, the highest bidder will usurp the second-highest bid and thereby win the auction, thus implementing the Vickrey outcome.

still ensuring a competitive market. But we have not yet developed a model, nor do we have evidence, to support this conclusion. Because of the bail-out option, however, we believe that two-stage auctions would encourage more bidders to enter the fray and, therefore, more competition – a sometimes overlooked consideration in the modeling of auctions (Rothkopf and Harstad, 1994).

[23] When there is only a single winner, such as for president, numerous social-choice criteria have been proposed, and their consequences investigated, for judging the quality of the winner (we will provide references in chapter 10). Because different voting procedures satisfy different criteria, it is perhaps not surprising that there is no consensus among social-choice theorists about a "best" system for electing a single winner.

Proof. Order the bids, b_i, of the n players such that $b_1 < b_2 < \ldots < b_n$. The proof proceeds by a series of five claims:

Claim 1. For $i = 1, 2, \ldots, n-2$, the bid b_i will not be usurped by those with bids $b_{i+1}, \ldots, b_n$.

Proof. We proceed by induction on i. For $i = 1$, notice that if anyone except the highest bidder in stage I usurps b_1 in stage II, then this will be a definite loss for that player, rendering this option less preferred than usurping a higher bid. CA thus prescribes that no one, except possibly the highest bidder, will usurp b_1. But now the highest bidder, realizing that no one else will usurp the lowest bid, knows that he or she will definitely lose if he or she usurps the lowest bid. Thus, the highest bidder will not usurp the lowest bid either. Hence, b_1 will not be usurped by anyone with a higher stage I bid.

Now assume that for each $i = 1, \ldots, k$, where $1 \leq k < n-2$, players know that b_i will not be usurped by anyone with a higher stage I bid. Because the highest bidder, in particular, will not usurp any of the bids $b_1, \ldots, b_k$, his or her stage II option will be at least b_{k+1}. Thus, if any other bidder usurps b_{k+1}, this bidder knows that he or she will definitely lose. Hence, CA again dictates that anyone with bids $b_{k+2}, \ldots, b_{n-1}$ will not usurp b_{k+1}. As before, the highest bidder, knowing this, will not usurp b_{k+1}, because this will be a definite loss. Thus, b_{k+1} will not be usurped by *anyone* with a higher stage I bid. This completes the induction.

Claim 2. Let H be the bidder with the highest valuation, h. If H bids h in stage I, then he or she will be the highest bidder in stage I and will usurp the second-highest bid in stage II.

Proof. Notice that if H bids h in stage I, by JBA his or her stage I bid will be the highest (i.e., b_n). Moreover, all H's options in stage II are ruled out by claim 1 except either to affirm b_n or usurp b_{n-1}. If H affirms b_n, then he or she receives a payoff equivalent to losing (i.e., H's profit is $b_n - h = 0$). Since usurping b_{n-1} is strictly better than this option, H will usurp b_{n-1}.

Claim 3. H can do no better than to win at the second-highest bid.

Proof. Let b denote the highest stage I bid among the players other than H. If H bids higher than b in stage I, by claim 1 the player who bid b will not usurp any stage I bid lower than b. Thus, H can do no better than to win by usurping b. If H bids lower than b in stage I, then the player who bid b is the highest bidder. This player will usurp no lower than the highest bid that is less than b, so H will have to usurp b (i.e., the highest bid) to win.

Claim 4. If a player is H, then his or her optimal strategy is to bid h and then usurp the second-highest bid.

Proof. Bidding $b_n = h$ and then usurping b_{n-1} yields a win for H at the second-highest bid. By claim 3, no strategy can yield a better result for H. By CA, then, H will bid h and then, by claim 2, usurp b_{n-1}.

Claim 5. For $j = 0, 1, \ldots, m$ (where m is the top possible valuation), if a player P has valuation $m - j$, then he or she will bid it. If P's bid is the highest, he or she will usurp the second-highest bid.

Proof. We proceed by induction on j. Suppose $j = 0$, so P has the highest possible valuation, m. Then P knows he or she is H, so claim 4 implies the desired conclusion. Now assume that $0 \leq k < m$, and we have shown that it is both true and common knowledge that

- if any P has valuation $m, m - 1, \ldots, m - k$, then he or she will bid it, whether or not P knows he or she has the highest valuation; and
- if P's bid turns out to be the highest stage I bid, he or she will usurp the second-highest bid.

Next assume that P has valuation $m - (k + 1)$. Then this is either the highest valuation, or it is not.

Case 1. P has the highest valuation.

In this case, claims 2 and 3 demonstrate that P can do no better than bid $m - (k + 1)$ in stage I. By JBA and CA, this will be P's stage I bid, and he or she will usurp the second-highest bid in stage II.

Case 2. P does not have the highest valuation.

In this case, we know, by our inductive hypothesis, that whoever has the highest valuation will bid it in stage I. Moreover, when that bid turns out to be the highest, this bidder (i.e., H) will usurp the second-highest bid in stage II and thereby win the auction. Thus, P with valuation $m - (k + 1)$ will definitely lose; by CA, P will therefore bid his or her valuation. In general, every P that is not H will also bid his or her valuation in stage I. (What P does in stage II we do not address here.)

This completes the proof of claim 5 and, with it, the theorem. Q.E.D.

10 Fair division by elections

10.1 Introduction

There are a host of voting procedures under which voters either can rank candidates in order of their preferences or allocate different numbers of votes to them, which we call *preferential voting systems* because they enable voters to distinguish more preferred from less preferred candidates. We shall describe four of the most common systems, discuss some properties that they satisfy, and illustrate paradoxes to which they are vulnerable. They are:

1 The Hare system of single transferable vote (STV)
2 The Borda count
3 Cumulative voting
4 Additional-member systems.

The rationale underlying all these systems is that of affording different factions or interests in the electorate the opportunity to gain representation in a legislature or council proportional to their numbers, which we call *proportional representation* (PR).[1] We shall briefly analyze each of these systems here and then make some comparisons, based on different criteria, at the end of the chapter. What is worth noting here, however, is that each of these systems offers a different approach to the problem of achieving PR, especially of minorities, which is the notion of fairness we take as our starting point in the study of elections with multiple winners.

The remainder of the chapter takes an unusual turn in that it was inspired by the request of a professional association to advise it on a voting procedure to use in electing its governing board. Although we shall not reveal the

[1] Other methods for achieving PR are analyzed in Rapoport, Felsenthal, and Maoz (1988a, 1988b). Paradoxes afflicting *list systems of PR* – in which voters can vote for one party, which receives representation in parliament proportional to its number of votes – are described in Van Deemen (1993). We do not analyze list systems here, because they achieve PR directly, at least to the degree that there is agreement on a method of rounding (see Balinski and Young, 1982, Young, 1994, chapter 3, and Ernst, 1994, for an analysis of different rounding methods).

association's identity, we have altered no essential facts of the study that was commissioned, which largely concerned how a fair division of seats on the governing board might be achieved through a new election procedure.

The association was reviewing its election procedures in response to pressures for reform. In particular, some members thought that certain types of members were underrepresented on the board – and other types over-represented – creating biases that affected the association's policies. A different voting system was seen as a possible way to address this perceived misrepresentation.

The association had, after considering each of the aforementioned PR systems, found it wanting in ensuring the kind of fair representation that it desired. For one thing, the leadership of the association did not want to encourage groups within the association to act like political parties in parliamentary systems, blatantly campaigning for seats to increase their shares of votes on the governing board. Indeed, the leadership viewed with repugnance the factionalism that such campaigns might induce in the association.

At the same time, the leadership wished to consider the possibility of a PR system that would smooth over existing divisions within the association by giving to aggrieved groups what they considered their fair share of seats on the board. The representation problem was complicated, however, by the fact that there was a desire not only that groups representing different specialties be represented but also that different regional interests be represented too (the association had offices worldwide).

After much discussion, the system that attracted the most interest was one designed to meet the specific needs of the association and dubbed "constrained approval voting" (CAV). Under CAV, the basic feature of "approval voting" (Brams and Fishburn, 1983) – that voters can vote for as many candidates as they like – is wedded to constraints placed on the numbers of candidates that can be elected in different categories. The winners under this system are the candidates most approved of by all the voters, subject to the constraints.

This hybrid system radically modified the purpose for which approval voting was originally designed. It also raised a number of questions about the properties of the constraints, and their likely effects, on the representation of different interests on the board. It is these that we shall discuss later in the chapter after first reviewing the four preferential systems mentioned earlier.[2]

[2] More extensive overviews of these systems can be found in, among other places, Dummett (1984), Nurmi (1987), and Merrill (1988). This and the next four sections are adapted from Brams and Fishburn (1991) and Brams (1994b) with permission.

10.2 The Hare system of single transferable vote (STV)

First proposed by Thomas Hare in England and Carl George Andrae in Denmark in the 1850s, *single transferable vote* (STV) is a voting system that involves the successive elimination of the lowest-vote candidates, and the transfer of the "surplus votes" of the highest vote-getters (once they have been elected), in a way we shall illustrate shortly. Despite its complexity, STV has been adopted throughout the world. It is used to elect public officials in such countries as Australia (where it is called the "alternative vote"), Malta, the Republic of Ireland, and Northern Ireland; in local elections in Cambridge, MA, and in local school board elections in New York City; and in numerous private organizations. John Stuart Mill (1862, p. 156) placed it "among the greatest improvements yet made in the theory and practice of government."

Although STV violates some desirable properties of voting systems (Kelly, 1987), it has strengths as a PR system. In particular, minorities can elect a number of candidates roughly proportional to their size in the electorate if they rank them high. Also, if one's vote does not help elect a first choice, it can still count for lower choices after higher choices are eliminated.

To describe how STV works and also to illustrate two properties that it fails to satisfy, consider the following examples (Brams, 1982; Brams and Fishburn, 1984c). The first shows that STV is vulnerable to "truncation of preferences" when two out of four candidates are to be elected, the second that it is vulnerable to "nonmonotonicity" when there is one candidate to be elected and there is no transfer of so-called surplus votes.

Example 1

Assume that there are three classes of voters, having a total of 17 votes, who rank the four candidates x, a, b, and c as follows:

I 6 voters: *xabc*
II 6 voters: *xbca*
III 5 voters: *xcab*.

Assume also that two from the set of four candidates are to be elected, and a candidate must receive a "quota" of six votes to be elected on any round. An *exact quota* is defined as $q = n/(m + 1) + 1$, where n is the number of voters (17 in our example) and m is the number of candidates to be elected (two in our example).

It is standard procedure to drop any fraction that results from the calculation of the exact quota, so what is actually used is the *quota*, $[q]$, which designates the integer part of q. The quota is the smallest integer that makes it impossible to elect more than m candidates by first-place votes on the first round. Since $[q]$ = six in our example ($q \cong 6.67$), at most two candidates can attain the quota on the first round (18 voters would be required for three candidates to get six first-place votes each). In fact, what happens is as follows:

1st round. x receives 17 out of 17 first-place votes and is elected.

2nd round. There is a *surplus* of 11 votes (i.e., the number greater than $[q] = 6$) that are transferred in the proportions 6:6:5 to the second choices (a, b, and c, respectively) of the three classes of voters. Since these transfers do not result in at least $[q] = 6$ for any of the remaining candidates (3.9, 3.9, and 3.2 votes for a, b, and c, respectively), the candidate with the fewest (transferred) votes (i.e., c) is eliminated under the rules of STV. The supporters of c (class III) transfer their 3.2 votes to their next-highest choice (i.e., a), giving a more than the quota (7.1 votes). Thus, a is the second candidate elected. Hence, the set of winners is $\{x, a\}$.

Now assume 2 of the 6 class II voters indicate x is their first choice, but they do not indicate a second or third choice. The new results are:

1st round. Same as earlier.

2nd round. There is a surplus of 11 votes (above $[q] = 6$) that are transferred in the proportions 6:4:2:5 to the second choices, if there are any (a, b, no second choice, and c, respectively), of the voters. (The 2 class II voters do not have their votes transferred to any of the remaining candidates because they indicated no second choice.) Since these transfers do not result in at least $[q] = 6$ for any of the remaining candidates (3.9, 2.6, and 3.2 votes for a, b, and c, respectively), the candidate with the fewest (transferred) votes (i.e., b) is eliminated. The supporters of b (four voters in class II) transfer their 2.6 votes to their next-highest choice (i.e., c), giving c 5.8 votes, less than the quota of six. Because a has fewer (transferred) votes (3.9), a is eliminated, and c is the second candidate elected. Hence, the set of winners is $\{x, c\}$.

Observe that the two class II voters who ranked only x first induced a better social choice for themselves by truncating their ballot ranking of candidates. Thus, it may be advantageous not to rank all candidates in order of preference on one's ballot, contrary to a claim made by a mathematical society that "there is no tactical advantage to be gained by marking few candidates" (Brams, 1982). Put another way, one may do better under the STV preferential system by *not* expressing preferences – at least beyond first choices – though it is doubtful that one would know this in advance.

The reason for this outcome in the example is that the two class II voters, by not ranking bca after x, prevent b's being paired against a (their last choice) on the 2nd round, wherein a beats b. Instead, c (their next to last choice) is paired against a and beats him or her, which is better for the class II voters.

Lest one think that an advantage gained by truncation requires the allocation of surplus votes, we next give an example in which only one candidate is to be elected, so the election procedure progressively eliminates candidates until one remaining candidate has a simple majority. This example illustrates a new and potentially more serious problem with STV than its manipulability due to preference truncation, which we shall illustrate first.

Example 2

Assume that there are four classes of voters, having a total of 21 votes, who rank the four candidates *a*, *b*, *c*, and *d* as follows:

I 7 voters: *abcd*
II 6 voters: *bacd*
III 5 voters: *cbad*
IV 3 voters: *dcba*.

Because no candidate has a simple majority of $[q] = 11$ first-place votes, the lowest first-choice candidate, *d*, is eliminated on the first round, and class IV's three second-place votes go to *c*, giving *c* eight votes. Because none of the remaining candidates has a majority at this point, *b*, with the new lowest total of six votes, is eliminated next, and *b*'s second-place votes go to *a*, who is elected with a total of 13 votes.

Now assume that the three class IV voters rank only *d* as their first choice. Then *d* is still eliminated on the first round, but since the class IV voters did not indicate a second choice, no votes are transferred. Now, however, *c* is the new lowest candidate, with five votes; *c*'s elimination results in the transfer of his or her supporters' votes to *b*, who is elected with 11 votes. Because the class IV voters prefer *b* to *a*, it is in their interest not to rank candidates below *d* to induce a better outcome for themselves, again illustrating the truncation problem.

It is true that under STV a first choice can never be hurt by ranking a second choice, a second choice by ranking a third choice, etc., because the higher choices are eliminated before the lower choice can affect them. However, lower choices can affect the order of elimination, and hence the transfer of votes. Consequently, a higher choice (e.g., second) can influence whether a lower choice (e.g., third or fourth) is elected.

We wish to make clear that we are not suggesting that voters would routinely make the strategic calculations implicit in these examples. These calculations are not only rather complex but also might be neutralized by counterstrategic calculations of other voters. Rather, we are saying that to rank all candidates for whom one has preferences is not always rational under STV, despite the fact that it is a ranking procedure.

Interestingly, STV's manipulability in this regard bears on its ability to elect *Condorcet candidates* (Fishburn and Brams, 1984) – those who can beat all other candidates in pairwise contests. Thus, in our example, *b* would beat *a* 14–7 (class II, III, and IV voters prefer *b* to *a*, whereas only class I voters prefer a to b); because b is also preferred by a majority of voters to both c and d, he or she is the unique Condorcet candidate in this example. But a's election can be upset in favor of b's if the three class IV voters, who have an interest in this upset's occurring, truncate their preferences.

Example 2 illustrates a potentially more serious problem with STV: raising a candidate in one's preference order can actually hurt that candidate, which is called *nonmonotonicity* (Smith, 1973; Doron and Kronick, 1977; Fishburn, 1982; Bolger, 1985). Thus, if the three class IV voters raise a from fourth to first place in their rankings – without changing the ordering of the other three candidates – b is elected rather than a. This is indeed perverse: a loses when he or she moves up in the rankings of some voters and thereby receives more first-place votes. Equally strange, candidates may be helped under STV if voters do not show up to vote for them at all, which has been called the *no-show paradox* (Fishburn and Brams, 1983; Ray, 1986; Moulin, 1988b; Holzman, 1988/89) and is related to the Condorcet properties of a voting system.

Because who is elected under STV – whether it is one candidate or more than one candidate – is vulnerable to voters' truncating their preferences, changing their rankings, or not showing up, STV's ability to give fair representation to the different classes of voters in our examples can be challenged. In particular, the fact that more first-place votes, or even no votes, can hurt rather than help a candidate violates what arguably is a fundamental democratic ethic. Before reaching any normative conclusions on this question, however, we shall assay the other preferential voting systems to see what, if any, properties they possess that may make the candidates they elect unresponsive to voter preferences.[3]

10.3 The Borda count

Under the *Borda count*, points are assigned to candidates such that the lowest-ranked candidate of each voter receives zero points, the next-lowest one point, and so on up to the highest-ranked candidate, who receives $m - 1$ votes if there are m candidates. Points for each candidate are summed across all voters, and

[3] Attempts to remedy certain problems of STV while retaining its basic features have been made. Most notable are systems proposed by Danny Kleinman, Single Office Majority Election (SOME) and Multiple Office Representative Election (MORE), in which candidates who are successively eliminated are those with the lowest Borda counts (see section 10.3). SOME guarantees the election of Condorcet candidates (illustrated in sections 10.2 and 10.3) if they exist, and MORE gives increased fractional weight to the ballots of those voters whose first choice is elected early, obviating the incentive of voters to rank shooins lower than they might otherwise when more than one candidate is to be elected (Kleinman, 1994). Kleinman has written much on voting and fair division over the years, but, unfortunately, his writings have not been published.

the candidate with the most points is the winner. To the best of our knowledge, the Borda count and similar scoring methods (Young, 1975) are not used to elect candidates in any public elections, but they are widely used in private organizations.

Like STV, the Borda count may not elect a Condorcet candidate, as illustrated by the case of three voters with preference order *abc*, and 2 voters with preference order *bca*. Under the Borda count, *a* receives six points, *b* seven points, and *c* two points, making *b* the Borda winner; yet *a* is the Condorcet candidate.

On the other hand, the Borda count would elect the Condorcet candidate (*b*) in example 2 in section 10.2. This is because *b* occupies the highest position on the average in the rankings of the four sets of voters, which range from best (1) to worst (4). Specifically, *b* ranks second in the ranking of 18 voters, third in the ranking of three voters, giving *b* an average ranking of 2.14, which is higher (i.e., closer to one) than *a*'s average ranking of 2.19 as well as the average rankings of *c* and *d*. Having the highest average ranking is indicative of being broadly acceptable to voters, unlike Condorcet candidate *a* in the example in the preceding paragraph, who is the last choice of two of the five voters.

The appeal of the Borda count in finding broadly acceptable candidates is similar to that of *approval voting*, under which voters can vote for, or approve of, as many candidates as they like but cannot rank them. Hence, it is not a preferential voting system that can ensure PR, so we will not consider it further here, even though it has several desirable properties in single-winner elections in which electing the most popular candidate – not achieving PR in a legislature – is the issue (Brams and Fishburn, 1983). But as noted in section 10.1, we will return to the subject of approval voting when we show how, in the incarnation of "constrained approval voting," it can be turned into a PR system.

Even in single-winner elections, the Borda count is almost certainly more subject to manipulation than approval voting. Consider again the example of the three voters with preference order *abc* and the two voters with preference order *bca*. Recognizing the vulnerability of their first choice, *a*, under the Borda count, the three *abc* voters might insincerely rank the candidates *acb*, maximizing the difference between their first choice (*a*) and *a*'s closest competitor, *b*. This would make *a* the winner under Borda.

In general, voters can gain under the Borda count by ranking the most serious rival of their favorite candidate last in order to lower his or her point total (Ludwin, 1978). This strategy is relatively easy to effectuate, unlike a manipulative strategy under STV that requires estimating who is likely to be eliminated, and in what order, so as to be able to exploit STV's dependence on sequential eliminations and transfers.

The vulnerability of the Borda count to manipulation led Jean-Charles de

Borda, after proposing his system in 1771, to exclaim, "My scheme is intended only for honest men!" (Black, 1958, pp. 182, 238). Nurmi (1984, 1987) has shown that the Borda count, like STV, is vulnerable to preference truncation, giving voters an incentive not to rank all candidates in certain situations. On the other hand, Chamberlin and Courant (1983) contend that the Borda count would give effective voice to different interests in a representative assembly, even if it would not always ensure their proportional representation with exactitude, and Saari (1994) argues it is more "determinate" than other systems.

Another type of paradox that afflicts the Borda count and related point-assignment systems involves manipulability by changing the agenda. For example, the introduction of a new candidate who cannot win – and, consequently, would appear irrelevant – can completely reverse the point-total order of the old candidates, even though there are no changes in the voters' rankings of these candidates (Fishburn, 1974b). Thus in the following example, the last-place finisher among three candidates (*a*, with six votes) jumps to first place (with 13 votes) when "irrelevant" candidate *x* is introduced, illustrating the extreme sensitivity of the Borda count to apparently irrelevant candidates:

3 voters: *cba*	3 voters: *cbax*
2 voters: *acb*	2 voters: *axcb*
2 voters: *bac*	2 voters: *baxc*
Outcome: $c = 8, b = 7, a = 6$	*Outcome:* $a = 13, b = 12, c = 11, x = 6$

It is worth noting in this example that there is no Condorcet candidate but, rather, *cyclical majorities*: $a > c > b > a$ – where "$>$" indicates that a majority of voters prefer the candidate to the left of the inequality over the candidate to the right of the inequality – so majority preferences are said to "cycle" (in this instance, from *a* back to *a*), indicating the lack of a social ordering along a single dimension. When every candidate can be defeated by another because majorities are cyclical, there is said to be a *paradox of voting*, which played a key role in Arrow's (1951) celebrated impossibility theorem.

Since there is no clear-cut social choice of the voters in such a situation, no candidate can be said to give them unequivocally fair representation. Candidate *c*, who had the highest average rank when *x* was not included, is reduced to having the lowest average rank of the three original candidates when *x* is added, because *x* pushes *c* farther down than the other original candidates.

To complicate matters further, assume that a second candidate, besides *a* in the election that includes *x*, were to be chosen to represent the voters. The Borda count says that candidate would be *b*, who received the second-highest Borda count, which would certainly please the two *baxc* voters. But should not the three *cbax* voters, because there are more of them, get their first-choice candidate next? It is apparent that the meaning of PR is suffused with ambiguity in a situation like this.

10.4 Cumulative voting

Cumulative voting is a voting system in which each voter is given a fixed number of votes to distribute among one or more candidates. This system allows voters to express their intensities of preference rather than simply to rank candidates, as under STV and the Borda count. It is a PR system in which minorities can ensure their approximate proportional representation by concentrating their votes on a subset of candidates commensurate with their size in the electorate.

To illustrate this system and the calculation of optimal strategies under it, assume there is a single minority position among the electorate favored by 1/3 of the voters, and a majority position favored by the remaining 2/3. Assume, further, that the electorate comprises 300 voters, and a six-member governing body is to be elected. The six candidates elected will be those with the greatest numbers of votes.

If each voter has six votes to cast for as many as six candidates, and if each of the 100 voters in the minority casts three votes each for only two candidates, these voters can ensure the election of these two candidates no matter what the 200 voters in the majority do. To see this, notice that each of the two minority candidates will get a total of 300 (100 x 3) votes, whereas the 2/3-majority, with a total of 1,200 (200 x 6) votes to allocate, can at best match this number for four candidates (1,200/4 = 300).

If the 2/3-majority instructs its supporters to distribute their votes equally among five candidates (1,200/5 = 240), it will not match the vote totals of the two minority candidates (300 each) but can still ensure the election of four (of its five) candidates – and possibly get its fifth candidate elected if the minority puts up three candidates and instructs its supporters to distribute their votes equally among the three (giving each 600/3 = 200 votes).

Against these strategies of either majority (support five candidates) or the minority (support two candidates), it is easy to show that neither side can improve its position. To elect five (instead of four) candidates with 301 votes each, the majority would need 1,505 instead of 1,200 votes, holding constant the 600 votes of the minority. Similarly, for the minority to elect three (instead of two) candidates with 241 votes each, it would need 723 instead of 600 votes, holding constant the 1,200 votes of the majority.

It is evident that the optimal strategy for the leaders of both the majority and minority is to instruct their members to allocate their votes as equally as possible among certain numbers of candidates. The number of candidates each supports should be approximately proportional to the number of their supporters in the electorate (if known).

Any deviation from this strategy – for example, by putting up a full slate of candidates and not instructing supporters to vote for only some on this slate –

offers the other side an opportunity to capture more than its proportional "share" of the seats. Patently, good planning and disciplined supporters are required to be effective under this system. In the face of uncertainty about their level of support in the electorate, party leaders may well make nonoptimal choices about how many candidates their supporters should concentrate their votes on, which weakens the argument that cumulative voting can in practice guarantee PR.

A systematic analysis of optimal strategies under cumulative voting is given in Brams (1975), including the role that uncertainty may play. These strategies are compared with strategies actually adopted by the Democratic and Republican parties in elections for the Illinois General Assembly, where cumulative voting was used until 1982.

Cumulative voting has also been used in elections for some boards of directors, which has enabled shareholders representing minority positions to gain seats on these boards. In 1987 cumulative voting was adopted by two cities in the United States (Alamogordo, NM, and Peoria, IL) – and other small cities more recently – to satisfy court requirements of minority representation in municipal elections. Cumulative voting has been most often used as a remedy in cities in which minority voters are widely scattered, rendering impractical the creation of so-called minority districts under a districting system.

10.5 Additional-member systems

In most parliamentary democracies, it is not candidates who run for office but political parties that put up lists of candidates. Under party-list voting, voters vote for the parties, which receive representation in a parliament proportional to the total numbers of votes that they receive. Usually there is a threshold, such as 5 percent, which a party must exceed in order to gain any seats in the parliament.

This is a rather straightforward means of ensuring the proportional representation of parties that attain or surpass the threshold. More interesting are *additional-member systems*, in which some legislators are elected from districts, but new members may be added to the legislature to ensure, insofar as possible, that the parties underrepresented on the basis of their national-vote proportions gain additional seats.

Denmark and Sweden, for example, use total votes, summed over each party's district candidates, as the basis for allocating additional seats. In elections to Germany's Bundestag and Iceland's Parliament, voters vote twice, once for district representatives and once for a party. Half of the Bundestag is chosen from party lists, on the basis of the national party vote, with adjustments made to the district results so as to ensure the approximate proportional representation of parties. In 1993 Italy adopted a similar system for its Chamber of Deputies, except that three-quarters rather than one-half of the seats are filled by deputies from the districts, with the remaining one-quarter used to approximate PR to the extent possible.

In Puerto Rico, no fixed number of seats is added unless the largest party in one house of its bicameral legislature wins more than two-thirds of the seats in district elections. When this happens, that house can be increased by as much as one-third to ameliorate the possible underrepresentation of minority parties.

To offer some insight into an important strategic feature of additional-member systems, assume, as in Puerto Rico, that additional members can be added to a legislature to adjust for underrepresentation, but this number is variable. More specifically, consider a voting system, called *adjusted districting voting*, or ADV (Brams and Fishburn, 1984a, 1984b), that has the following attributes:

1 There is a jurisdiction divided into equal-size districts, each of which elects a single representative to a legislature.
2 There are two main factions in the jurisdiction, one majority and one minority, whose size can be determined. For example, if the factions are represented by political parties, their respective sizes can be determined by the votes that each party's candidates, summed across all districts, receive in the jurisdiction.
3 The legislature consists of all representatives who win in the districts *plus* the largest vote-getters among the losers – necessary to achieve PR – if it is not realized in the district elections. Typically, this adjustment will involve adding minority-faction candidates, who lose in the district races, to the legislature so that it mirrors the majority-minority breakdown in the electorate as closely as possible.
4 The size of the legislature is *variable*, with a lower bound equal to the number of districts (if no adjustment is necessary to achieve PR), and an upper bound equal to twice the number of districts (if a nearly 50 percent minority wins no seats).

As an example of ADV, suppose that there are eight districts in a jurisdiction. If there is an 80 percent majority and a 20 percent minority, the majority is likely to win all the seats unless there is an extreme concentration of the minority in one or two districts.

Suppose the minority wins no seats. Then its two biggest vote-getters could be given two "extra" seats to provide it with representation of 20 percent in a body of ten members, exactly its proportion in the electorate.

Now suppose the minority wins one seat, which would provide it with representation of $1/8 \cong 12$ percent. If it were given an extra seat, its representation would rise to $2/9 \cong 22$ percent, which would be closer to its 20 percent proportion in the electorate. However, assume that the addition of extra seats can never make the minority's proportion in the legislature exceed its proportion in the electorate, which we call the *proportionality constraint*.

Paradoxically, the minority would benefit by winning no seats and then being granted two extra seats to bring its proportion up to exactly 20 percent. To prevent a minority from benefiting by *losing* in district elections, assume the following *no-benefit constraint*: the allocation of extra seats to the minority can never give it a greater proportion in the legislature than it would obtain had it won more district elections.

How would this constraint work in our example? If the minority won no seats in the district elections, then the addition of two extra seats would give it $2/10 = 1/5$ representation in the legislature, exactly its proportion in the electorate. But we just showed that if the minority had won one seat, it would not be entitled to an extra seat – and 2/9 representation in the legislature – because of the proportionality constraint. Hence, its representation would have to remain at 1/8 if it won in exactly one district.

Because $1/5 > 1/8$, the no-benefit constraint prevents the minority from gaining two extra seats if it wins no district seats initially. Instead, it would be entitled in this case to only one extra seat, giving it 1/9 rather than 1/5 representation; since $1/9 < 1/8$, the no-benefit constraint is satisfied.

But $1/9 \cong 11$ percent is only about half the minority's 20 percent proportion in the electorate. In fact, one can prove in the general case that the no-benefit constraint may prevent a minority from receiving up to about half of the extra seats it would be entitled to – on the basis of its national-vote total – were the constraint not operative and it could therefore get up to this proportion (e.g., two out of ten seats in our example) in the legislature (Brams and Fishburn, 1984a).

The no-benefit constraint may be interpreted as a kind of "strategyproofness" feature of ADV: it makes it unprofitable for a minority party deliberately to lose in a district election in order to do better after the adjustment that gives it extra seats. But strategyproofness, in precluding any possible advantage that might accrue to the minority from throwing a district election, has a price.

As our example demonstrates, it may severely restrict the ability of ADV to satisfy PR (11 percent versus 20 percent for the minority), giving rise to the following dilemma: Under ADV, one cannot guarantee PR if one wishes to satisfy both the proportionality constraint and the no-benefit constraint, which allows the minority only 11 percent representation in our example. But dropping the no-benefit constraint allows the minority to obtain its full proportional representation (two out of ten seats in the augmented legislature). Yet this gives the minority an incentive deliberately to lose in the district contests in order to do better after the adjustment (i.e., get its full 20 percent by winning no district seats).

It is worth pointing out that the "second chance" for minority candidates afforded by ADV would encourage them to run in the first place, because even if most or all of them are defeated in the district races, their biggest vote-getters would still get a chance at the (possibly) extra seats in the second stage. But these extra seats might be cut by a factor of up to two from the minority's proportion in the electorate, assuming one wants to motivate district races with the no-benefit constraint.

Consider what might have happened under ADV after the June 1983 British general election. The Alliance (comprising the new Social Democratic party and the old Liberal party) got more than 25 percent of the national vote but less than 4 percent of the seats in Parliament (by coming in second in most districts behind the Conservative or Labour party candidates). ADV would have assigned the Alliance 20 percent representation in an augmented 817-seat Parliament – 26 percent larger than the present 650-seat Parliament[4] – which is somewhat less than the Alliance's proportional 25 percent share that the no-benefit constraint precludes (Brams and Fishburn, 1984b).

In the remainder of the chapter, we offer a detailed analysis of "constrained approval voting" (CAV), the new election procedure we mentioned in section 10.1 that was designed to meet the requirements of a professional association for the election of its governing board. We begin by providing some background on the association and describing how one finds "controlled roundings."

10.6 Controlled roundings

In previous elections, association members voted for a slate of candidates prepared by a nominating committee.[5] The committee prepared a slate of about twice as many candidates as there were seats to be filled; members could vote only for as many candidates as there were seats to be filled – no more and no less. Those candidates with the most votes were elected to fill the open seats over a multiyear cycle. If the cycle were two years, for example, board members would serve two years, with half the seats being filled in each annual election.

No change was contemplated in the size of the board or in the terms of office of its members. The issue was whether to elect board members by constituencies representing different interests of the association. What made the problem unusual was that a constituency was defined by two dimensions – region and specialty – rendering the problem of assigning seat shares more

[4] Under ADV, it might be advisable to cut considerably the number of districts in Britain so that, after adjustments are made, Parliament would generally be close to its present size and therefore not too unwieldy.

[5] This and the next two sections are adapted from Brams (1990a) with permission.

difficult than if only one dimension (e.g., a left–right policy continuum, along which political parties may take positions) needed to be represented.[6]

To illustrate the two-dimensional problem, assume that there are two *regional* divisions of the association (A and B) and three *specialty* divisions (X, Y, and Z). The percentages of members that fall into each category can be shown in a 2 x 3 matrix, where the rows indicate the regional divisions and the columns the specialty divisions:

	Specialty			
Region	X	Y	Z	Row total
A	27	16	17	60
B	21	9	10	40
Column total	48	25	27	100

These percentages be interpreted as targets for the composition of the board.

The targets in any election will depend not only on the numbers of members that fall into each category but also on the numbers of board members continuing in each category. For example, if elections are held over a two-year cycle, and cell AY already contains 24 percent of the continuing board members, the 16 percent shown in the table should be reduced to 8 percent as a target – and underrepresented cells increased accordingly – to ensure that members from AY do not continue to be overrepresented but instead are properly represented at the 16 percent level (the average of 24 percent and 8 percent) on the next board.

The percentages shown in the table are an ideal: given that each board member has one vote, no allocation of board members (and therefore votes) to each category will mirror the percentages perfectly. Although a system of weighted voting could lead to a better fit, no consideration was given to endowing different board members with different numbers of votes.

For the next step, assume that the percentages in the matrix are the targets, and six new members are to be elected to the board. (By coincidence, this number exactly matches the number of cells.) Multiplying the target percentages by 6, we obtain the following target election figures, or TEFs:

[6] The mathematical problem, nonetheless, of assigning an integer number of seats in a parliament to parties, based on the votes they receive, is not a trivial one. Balinski and Young (1982) offer a thorough analysis of this problem – mainly in the context of assigning seats to congressional districts based on their populations – and also provide an engrossing history of this problem; for a critical discussion of their proposed solution to the one-dimensional apportionment problem, see Brams and Straffin (1982); see also Woodall (1986a) and Young (1994, chapter 3).

Region	Specialty X	Y	Z	Row total
A	1.62	0.96	1.02	3.60
B	1.26	0.54	0.60	2.40
Column total	2.88	1.50	1.62	6.00

The remainders of each of the TEFs preclude a perfect matching of (whole) representatives to the cells. But this does not prevent us from narrowing the possibilities to those that are, in some sense, best-fitting. To do so, consider the following set of constraints:

(1) *Row and column minima.* Rounded down, the TEF column sums are 2, l, and l; the row sums are 3 and 2. Assuming these as *minima* for the totals of regional and specialty representatives, respectively, there are 65 distinct cases that satisfy these constraints and whose cell entries sum to 6, as exhaustively enumerated in table 10.1.

There are systematic procedures but seem to be no efficient algorithms for generating all these cases. For a 2 x 3 matrix, a hand calculation is feasible; for larger matrices, the association used computer spreadsheets to find integer allocations, reflecting finer breakdowns of the association.

(2) *Row and column maxima.* Rounded up, the TEF column sums are 3, 2, and 2; the row sums are 4 and 3. Assuming these as *maxima* for the totals of regional and specialty representatives, respectively, 30 of the 65 cases satisfying constraint l are excluded. Specifically, the row maxima exclude none of the 65 cases, but the first, second, and third column maxima exclude 8, 11, and 11 cases, as shown in table 10.2.

These column-maxima constraints are mutually exclusive: the cases excluded by each one are not excluded by either of the other two. Thirty-five cases remain admissible.

(3) *Cell minima and maxima.* Rounding down and up the TEFs of all the cells gives a minimum and a maximum for each cell. Satisfying these minimal and maximal cell constraints reduces the 35 admissible cases meeting constraints l and 2 to just ten "controlled roundings," as shown in table 10.3. Note that the sum of the first row in the first five cases is 4, whereas this sum in the last five cases is 3.

The three constraints, applied progressively, have reduced the number of admissible cases from 65 (constraint l) to 35 (constraints l and 2) to ten (constraints l, 2, and 3) whose cell entries, and column and row totals, sum to

Table 10.1. *65 cases that satisfy criterion 1: row and column sums are no less than the row and column TEFs rounded down (minima)*

Sum of first row = 4 (30 cases)								
4 0 0 0 1 1	3 1 0 1 0 1	3 1 0 0 1 1	3 1 0 0 0 2	3 0 1 1 1 0	3 0 1 0 2 0	3 0 1 0 1 1	2 2 0 1 0 1	2 2 0 0 1 1
2 2 0 0 0 2	2 1 1 2 0 0	2 1 1 1 1 0	2 1 1 1 0 1	2 1 1 0 2 0	2 1 1 0 1 1	2 1 1 0 0 2	2 0 2 1 1 0	2 0 2 0 2 0
2 0 2 0 1 1	1 3 0 1 0 1	1 2 1 2 0 0	1 2 1 1 1 0	1 2 1 1 0 1	1 1 2 2 0 0	1 1 2 1 1 0	1 1 2 1 0 1	1 0 3 1 1 0
0 3 1 2 0 0	0 2 2 2 0 0	0 1 3 2 0 0						
Sum of first row = 3 (35 cases)								
3 0 0 1 1 1	3 0 0 0 2 1	3 0 0 0 1 2	2 1 0 2 0 1	2 1 0 1 1 1	2 1 0 1 0 2	2 1 0 0 2 1	2 1 0 0 1 2	2 1 0 0 0 3
2 0 1 2 1 0	2 0 1 1 2 0	2 0 1 1 1 1	2 0 1 0 3 0	2 0 1 0 2 1	2 0 1 0 1 2	1 2 0 2 0 1	1 2 0 1 1 1	1 2 0 1 0 2
1 1 1 3 0 0	1 1 1 2 1 0	1 1 1 2 0 1	1 1 1 1 2 0	1 1 1 1 1 1	1 1 1 1 0 2	1 0 2 2 1 0	1 0 2 1 2 0	1 0 2 1 1 1
0 3 0 2 0 1	0 2 1 3 0 0	0 2 1 2 1 0	0 2 1 2 0 1	0 1 2 3 0 0	0 1 2 2 1 0	0 1 2 2 0 1	0 0 3 2 1 0	

the grand total of 6. Satisfying these three constraints results in what is called a *controlled rounding*, which can always be found for any matrix (Cox and Ernst, 1982). For larger arrays (i.e., three dimensions or more), however, a controlled rounding may not exist (Fagan, Greenberg, and Hemmig, 1988).

A controlled rounding can be defined more straightforwardly as one in which, for every column and row, the sum of its cell TEFs, rounded down or up, equals the column or row (total) TEF, rounded down or up, with the roundings summing to some grand total. Constraints 1 and 2 give all possible cases that are roundings of the column and row TEFs and sum to 6; constraint 3 limits these to those that are also roundings of the cell TEFs.

10.7 Further narrowing: the search may be futile

One could reduce the number of cases still further by invoking various criteria. For example, define an integer representation to be *cell-consistent* if the TEF of a cell that is assigned a larger integer is at least as great as one that is assigned a smaller integer. In controlled-rounding case #1, the TEF of cell BY is 0.54

Table 10.2. *65 cases that satisfy criterion 1 and are either excluded or included by criterion 2: row and column sums are no more than the row and column TEFs rounded up (maxima)*

30 cases excluded

8 cases in which the first column sums to more than 3

4 0 0	3 1 0	3 0 1	2 1 1	3 0 0	2 1 0	2 0 1	1 1 1
0 1 1	1 0 1	1 1 0	2 0 0	1 1 1	2 0 1	2 1 0	3 0 0

11 cases in which the second column sums to more than 2

0 2 2	2 1 1	1 3 0	1 2 1	0 3 1	2 1 0	2 0 1	1 2 0	1 1 1
0 1 1	0 2 0	1 0 1	1 1 0	2 0 0	0 2 1	0 3 0	1 1 1	1 2 0
0 3 0	0 2 1							
2 0 1	2 1 0							

11 cases in which the third column sums to more than 2

2 1 1	2 0 2	1 1 2	1 0 3	0 1 3	2 1 0	2 0 1	1 1 1	1 0 2
0 0 2	0 1 1	1 0 1	1 1 0	2 0 0	0 0 3	0 1 2	1 0 2	1 1 1
0 1 2	0 0 3							
2 0 1	2 1 0							

35 cases included

16 cases in which the first row sums to 4

3 1 1	3 1 0	3 0 1	3 0 1	2 2 0	2 2 0	2 1 1	2 1 1	2 1 1
0 1 1	0 0 2	0 2 0	0 1 1	1 0 1	0 0 2	1 1 0	1 0 1	0 1 0
2 0 2	2 0 2	1 2 1	1 2 1	1 1 2	1 1 2	0 2 2		
1 1 0	0 2 0	2 0 0	1 0 1	2 0 0	1 1 0	2 0 0		

19 cases in which the first row sums to 3

3 0 0	3 0 0	2 1 0	2 1 0	2 1 0	2 0 1	2 0 1	2 0 1	1 2 0
0 2 1	0 1 2	1 1 1	1 0 2	0 1 2	1 2 0	1 1 1	0 2 1	2 0 1
1 2 0	1 1 1	1 1 1	1 1 1	1 0 2	1 0 2	0 2 1	0 2 1	0 1 2
1 0 2	2 1 0	2 0 1	1 1 1	2 1 0	1 2 0	3 0 0	2 0 1	3 0 0
0 1 2								
2 1 0								

and that of BZ is 0.60; yet BY is assigned a 1 and BZ a 0, which makes this representation cell-inconsistent.

In fact, the only two cases that are cell-consistent are #2 and #9. Allocation #2 is cell-consistent because cell AX is the largest TEF (1.62) and receives the only two seats that are assigned to a cell; BY is the smallest TEF (0.54) and receives the only zero. Allocation #9 is cell-consistent because l's are assigned

Table 10.3. *10 controlled roundings that satisfy criteria 1 and 2 and, in addition, criterion 3: cell entries are no less than the cell TEFs rounded down (minima), and no more than the cell TEFs rounded up (maxima)*

#1	#2	#3	#4	#5	#6	#7	#8	#9	#10
211 110	211 101	112 200	112 110	202 110	201 111	111 210	111 201	111 111	102 210

to all cells, so no smaller TEF receives a larger assignment than a larger TEF. When the integer assignments to the column and row sums are also consistent, the allocation is said to be *consistent.*

Another criterion for reducing the number of integer representations – in this case, to exactly one – is the Hamilton method of rounding (Balinski and Young, 1982), which has two steps:

1 Allocate to each category – both the six cells and the column and row sums – the integer portion of its TEF (i.e., its number to the left of the decimal point);
2 Of those seats remaining (out of the six to be allocated in the example), allocate them to the TEFs with the largest remainders – starting with the TEF with the biggest remainder – until the six seats are exhausted.

To illustrate the Hamilton method for the TEFs given earlier, the integer allocations according to step 1 are as follows (note that the column sums total 4 and the row sums total 5):

```
| 1  0  1  3
| 1  0  0  2

  2  1  1  4/5
```

The remaining seats are now allocated, according to step 2, on the basis of the TEFs having the largest remainders:

- three seats to cells in the 2 x 3 matrix (to which three seats have already been allocated);
- two seats to the column sums (to which four seats have already been allocated);
- one seat to the row sums (to which five seats have already been allocated).

These assignments give as a final allocation

```
| 2  1  1  4
| 1  0  1  2

  3  1  2  6
```

This allocation, called a Hamilton allocation, is the same as (consistent) allocation #2 in table 10.3. Allocation #9, on the other hand, is consistent but not Hamilton, which illustrates

Proposition 10.1 Hamilton allocations are always consistent, but consistent allocations are not always Hamilton.

Proof. The second part of this proposition is proved by allocation #9. The first part follows from the fact that, by step 1 of the Hamilton method, TEFs with larger integer portions never receive fewer seats than TEFs with smaller integer portions; by step 2, TEFs with larger remainders never receive fewer seats than TEFs with smaller remainders. Hence, larger TEFs can never be assigned fewer seats than smaller TEFs. Q.E.D.

Controlled roundings #3 and #5, in addition to the Hamilton allocation (#2), have column and row sums identical to that of the Hamilton allocation. However, these allocations are cell-inconsistent and hence inconsistent. By Proposition 10.1, they cannot be Hamilton, because Hamilton allocations are a subset of consistent allocations. (Except for possible ties, in which a seat might be randomly assigned at step 1 or step 2, Hamilton allocations are unique.)

So far, it would appear, a Hamilton allocation is the most sensible of the (consistent) controlled rounding allocations. But there is a rub: one may not exist.

For example, the following percentages and TEFs for the allocation of six seats differ very little from those used in the earlier example:

Percentages

30	10	19	59
18	12	11	41
48	22	30	100

Target election figures

1.80	0.60	1.14	3.54
1.08	0.72	0.66	2.46
2.88	1.32	1.80	6.00

Now applying the Hamilton method to the TEFs – both the cells and the column and row sums – one obtains the following allocations for each:

2	0	1	4
1	1	1	2
3	1	2	6

Although the column allocations sum to their Hamilton allocations, the first row sums to 3 (not 4), and the second row sums to 3 (not 2). This example proves that Hamilton allocations to the cells may not agree with Hamilton allocations to the column or row sums.

The fact that the Hamilton allocations in this example are unique, but not a controlled rounding, immediately implies

Proposition 10.2 A controlled rounding that is Hamilton may not exist.

A requirement of a controlled rounding is that the sums of the rounded cell TEFs for each column and row equal the corresponding rounded column and row TEFs.

Finally, to settle the question of the existence of a consistent controlled rounding, consider the following percentages for a 2 x 2 matrix and, supposing three seats are to be filled, the TEFs:

Percentages

30	22	52
21	27	48
51	49	100

Target election figures

0.90	0.66	1.56
0.63	0.81	1.44
1.53	1.47	3.00

The only cell-consistent allocation is to assign one seat to all entries except the lowest (0.63). But the consistency of the column sums demands that the first column receive two seats, when in fact the sum of its cell-consistent entries (0 + l) is l. This example proves

Proposition 10.3 A controlled rounding that is consistent may not exist.

There is a final difficulty with cell-consistent controlled roundings, illustrated by the following percentages and TEFs:

Percentages

14.6	9.4	18.0	42.0
9.6	9.6	0.4	19.6
19.8	9.2	9.4	38.4
44.0	28.2	27.8	100.0

Target election figures

0.73	0.47	0.90	2.10
0.48	0.48	0.02	0.98
0.99	0.46	0.47	1.92
2.20	1.41	1.39	5.00

The cell-consistent allocation of seats shown below is not consistent because the first two columns and the second two rows do not sum to values consistent with their column and row TEFs:

1	0	1	2
1	1	0	1
1	0	0	2
2	2	1	5

However, a new difficulty arises in this example: the second row is entitled to only 0.98 seats, but the cell-consistent assignments for this row sum to 2.

When the discrepancy between the cell-consistent sum of a column or row and its TEF is greater than 1.0, it does not satisfy *quota* (Balinski and Young, 1982). Put another way, the column or row TEF, rounded either up or down, is not equal to the cell-consistent sum. In the example, 0.98 rounded up is 1, but the cell-consistent sum of the second row is 2, which proves

Proposition 10.4 A cell-consistent rounding may not satisfy quota.

An assignment of seats that violates quota, of course, is not a controlled rounding.

Because the narrowing-down criteria we have discussed – consistent allocations and Hamilton allocations – can be incompatible with all controlled roundings, they cannot reliably be used to distinguish either a very few or a single best allocation. In addition, a cell-consistent allocation not only can be inconsistent but also can fail to satisfy quota – and therefore not be a controlled rounding.

Other criteria have been proposed for filtering out the best-fitting controlled roundings. For example, Balinski and Demange (1989) and Gassner (1991) provide excellent analyses of technical criteria for finding best-fitting biproportional allocations (that is, those proportional to the TEFs in two dimensions, as here); they even suggest their application in a political context, where biproportionality might be based on political parties and geographical constituencies rather than specialties and regions. Cox (1987) argues for unbiased controlled roundings and provides a computationally efficient procedure for finding them. Other approaches and algorithms have been proposed (Kelly, Golden, and Assad, 1989), but they are not tied, in our view, to fundamental principles of fair representation.

For the purpose of choosing an elected board that is a reasonable approximation of the TEFs, the fact that no controlled rounding may satisfy a requirement as weak as consistency – not to mention give allocations compatible with a specific apportionment method like Hamilton – casts doubt on the recommendation of one allocation on purely theoretical grounds. Indeed, a case can be made that any of the controlled roundings is good enough: each cell receives representation within one seat of what it is entitled to, and so does each geographical region and functional category.

In the absence of compelling criteria for singling out a best controlled rounding in an election, an empirical solution was recommended to the association. Underlying it, however, was a theoretical rationale tied to the notion of voter sovereignty. This was to let the voters themselves choose the outcome they most favored.

10.8 Constrained approval voting (CAV)

As a starting point, it was suggested that the association use approval voting (sections 10.1 and 10.3). Under approval voting, members would be able to vote for as many candidates as they approved of, or found acceptable, rather than – as under the extant system – be restricted to voting for exactly as many candidates as there were seats to be filled. But, as before, members could still vote for any candidates, irrespective of their regional or specialty designation, which now would be explicitly indicated on the ballot.

The advantages of approval voting have been discussed in detail elsewhere (Brams and Fishburn, 1983; Nurmi, 1987; Merrill, 1988). However, these advantages pertain mainly to single-winner elections with more than two candidates, for which the proportional representation of different kinds of representatives on a board is not the question we consider here.

Still, there seemed no good reason to force voters to vote for an arbitrary number of candidates. Rather, it was argued, they should be permitted the more flexible option of voting for as many candidates as they liked.

The flexibility afforded by approval voting made sense to association members, who testified that many members did not have sufficient knowledge to make more than two or three informed choices. Thus, the requirement that they vote for, say, six candidates forced less knowledgeable members (usually newer) to make less-than-informed judgments, often influenced by casual advice from more senior members.

The second recommendation – to restrict the domain of possible outcomes – was designed to counteract a possible bias that approval voting might introduce. Specifically, if members of the largest categories, like AX with 27 percent of the members in our earlier example, tended to concentrate their votes on candidates in this category, they could unduly affect the election outcome.

In fact, even under the extant system, in which voters were required to vote for an entire slate, the nominating committee "engineered" the slate to thwart voters from electing "too many" members of one type. (This was referred to as *slate engineering*, which might be roughly defined as rigging an election to produce certain desired results.) For example, if AX members were over-represented on the continuing board, the nominating committee might propose AX candidates who were not well known in order to diminish their chances of election.

This form of manipulation is well known to political scientists (Riker, 1986). But it is an informal device that on occasion did not work as planned, which is one reason why the association wanted to explore alternatives that offered more formal protection. Presumably, an election system that explicitly ensured the fair representation of different interests would also gain the confidence of voters. Consequently, whatever outcome it produced would be considered more legitimate.

If the admissible outcomes are restricted to the set of controlled roundings and not a particular one, then voters would decide not only who is elected from each cell but also, within limits, how many. Of course, limiting voters to outcomes in the set of controlled roundings is drastically different from using approval voting to elect single winners in multicandidate elections, with no restrictions on who can be elected.

Indeed, one might argue that the ten controlled roundings in the earlier example are too restrictive a set. The 35 (or even 65) cases available if criterion 3 (or 2 as well) were lifted would give the voters more control in the choice of a board and hence greater sovereignty.

Thereby, the board's composition would be more responsive to their voting. Whereas the controlled roundings guarantee that an integer representation is no more than one seat from the TEFs, the less restrictive set of, say, 35 cases would permit 25 additional outcomes – each of which leads to the election of at least one different candidate – and still guarantee column and row sums within one seat of their TEFs.

The acceptability of this set versus the ten controlled roundings depends upon the importance one attaches to the principle that the number of cell seats – versus the regional and specialty totals – should all be within one seat of the TEFs. To put this matter somewhat differently, the designation of what outcomes are admissible will depend on whether it is thought that more popular (i.e., approved) candidates should be permitted to win at the price of causing deviations from the cell TEFs greater than one seat.

Once the voters have chosen the set of outcomes they deem admissible, the outcome they select under CAV will be that with the greatest total number of approval votes. For example, if the admissible outcomes are the ten controlled roundings in the earlier example, they have in common the certain election of exactly one candidate from the three cells AX, AZ, and BX:

1	0	1
1	0	0

This means that the biggest vote-getters in each of these cells will be guaranteed election, whichever of the ten controlled roundings wins. The votes for these candidates can then be set aside.

The particular controlled rounding that wins will be the one in which the total approval vote for the three remaining "discretionary" choices is greatest. In case #1, for example, the runner-up in cell AX, the winner in cell AY, and the winner in cell BY would complete the six choices. The total of the votes for these three candidates would be compared with analogous totals in the nine other cases – given by the sums of the votes of the three best-performing candidates in the appropriate cells who were not certain winners – to determine the winning controlled rounding.[7]

If one admits the 35 cases that meet criteria 1 and 2, what they have in common is that no candidates are guaranteed election from any cell (see figure 10.2). In this set of cases, all six choices are discretionary, although the constraints that these cases – as well as the ten controlled roundings – must satisfy still impose restrictions on the possible outcomes (e.g., four discretionary choices cannot all be chosen from one cell). At the least stringent level of criterion 1 alone, the election of four candidates from cell AX is permitted in one case (see figure 10.1), which is more than a two-seat (122 percent) deviation from its TEF of 1.80.

In our opinion, this deviation is too large, and we therefore recommend tighter restrictions. Both the ten controlled roundings and the 35 cases that ensure the row and column totals will be within one seat of the TEFs seem acceptable to us. Making this choice, the association could determine how much leeway it wants to permit the voters.

Clearly, approval voting, by allowing voters to vote for as many candidates as they like, gives voters greater sovereignty than does restricting their votes to a fixed number. But where one draws the line to preclude the election of candidates who would not form a representative board is a value judgment. Our analysis, we believe, clarifies the tradeoffs that that judgment entails.

[7] Integer programs that efficiently accomplish this calculation are given in Potthoff (1990) and Straszak *et al.* (1993).

10.9 Conclusions

There is no perfect voting procedure. Because different procedures satisfy different desiderata, one must decide on those desiderata one considers most important in order to make a choice.[8] In this chapter, we focused on ranking systems and a new system, constrained approval voting (CAV), that are designed to ensure PR, which we took to be a measure of fair division in a legislature or council.

We began the analysis with single transferable vote (STV), a rather complicated successive-elimination and transfer-of-surplus-points procedure that has been used in both public elections in several countries and private elections in numerous organizations. It elects all candidates who surpass a certain quota after the eliminations and transfers.

STV's vulnerability to preference truncation and the no-show paradox illustrates its manipulability, and its nonmonotonicity casts doubt upon its democratic character. In particular, it seems bizarre that voters, because of nonmonotonicity, may hurt candidates – even prevent them from winning – by raising them in their rankings.

The Borda count is a scoring method, under which voters give more points to their higher-ranked candidates. The candidates with the most points win. Although the Borda count is monotonic, it is easier to manipulate than STV, which, because of its complexity, renders strategic calculations difficult. By contrast, the strategy of ranking the most serious opponent of one's favorite candidate last under the Borda count is an obvious way to diminish the opponent's chances. The Borda count's vulnerability to manipulation is further underscored by the fact that the introduction of a new and seemingly irrelevant candidate can have a topsy-turvy effect on the election results, moving a last-place candidate into first place and vice versa.

Cumulative voting offers a means for parties (or other groups) to guarantee their proportional representation, whatever the strategies of other parties, by concentrating their multiple votes on a number of candidates proportional to the party's size in the electorate. However, its effective use requires considerable organizational effort on the part of parties – and the disciplined behavior

[8] There are further analogies between some well-known properties of voting systems and properties, such as envy-freeness and efficiency, in the fair-division context. For example, both envy-freeness and efficiency can be described in terms of the desirability of trading: an allocation is envy-free if no player desires to trade his or her whole share for someone else's whole share; it is efficient if no sequence of trades of parts of players' allocations can improve everyone's lot. In a similar vein, the question of whether or not a voting system is "weighted" turns out to be equivalent to asking if there are some losing coalitions, all of which could gain (i.e., become winning) by a trade of players (Taylor and Zwicker, 1992). More in this vein can be found in Taylor and Zwicker (1993, 1995a, 1995b, 1995c) and Taylor (1995).

of their members – especially in the face of uncertainty about the parties' level of support in the electorate before the election.

Additional-member systems, and specifically adjusted district voting (ADV) that results in a variable-size legislature, provide a mechanism for approximating PR in a legislature without the nonmonotonicity of STV, the manipulability of the Borda count, or the organizational efforts or discipline required of cumulative voting. But imposing the no-benefit constraint on the allocation of additional seats – in order to eliminate the incentive of parties to throw district races – also vitiates fully satisfying PR, underscoring the difficulties of satisfying a number of desiderata. An understanding of these difficulties, and the possible tradeoffs they entail, facilitates the selection of a PR procedure to meet the needs one considers most pressing.

We return to the case of constrained approval voting (CAV) and what the professional association decided about its use. A majority of the committee that had been formed to consider election reform proposals – and make recommendations to the board – thought that breaking the association down into regional and specialty categories would violate its unitary philosophy. Instead, they wanted its members to view it as a single entity.

Consequently, the association decided to continue to use slate engineering to ensure, insofar as possible, a representative board. And once it had rejected categorizing candidates, it saw approval voting as a secondary issue and did not deem it desirable without the constraints.

The failure of the professional association to adopt CAV seems also to have been a function of another factor. Most of the work done on the design of CAV was with a subcommittee of about five people, but the full committee that made the final decision on adoption had about ten additional members. Although these people received extensive written reports summarizing the work of the subcommittee, they probably did not fully appreciate CAV's advantages over slate engineering, which had worked fairly well in the past.

The options (including CAV) presented to the full committee when it convened were phrased neutrally; moreover, ample time was given to discuss the arguments for CAV. However, despite rumblings of discontent about the make-up of the board, there was no crisis at the time. Furthermore, because the committee had previously considered and rejected other voting systems, including the ranking systems that we analyzed earlier, most members felt that they had given due consideration to possible alternatives. In short, there seemed no overriding reason to make a change.

By comparison, professional societies that have adopted approval voting (without the constraints) have done so, in most cases, in elections with a single winner (e.g., for president). They have generally done so because approval voting appeared to be manifestly superior to alternative systems, like *plurality voting* – voters can vote for only one candidate, and the candidate with the most votes wins – in finding a consensus choice in multicandidate races. These societies have included The Institute of Management Sciences (TIMS, with about 8,000 members), which in 1995 merged with the Operations Research Society of America (ORSA, also with about 8,000 members) to form the Institute for Operations Research and the Management Sciences (INFORMS) that now has used approval voting, the Mathematical Association of America (MAA, with about 32,000 members), the American Statistical Association (ASA, with about 15,000 members), and the Institute of Electrical and Electronics Engineers (IEEE, with about 350,000 members).

Members of these societies, some of whom were already familiar with approval voting, were instrumental in its adoption in their societies. In addition, academics who had done research on approval voting played a role in most of the adoption decisions (Brams and Fishburn, 1992).

These connections led to several empirical analyses of election returns of these societies (Brams, 1988; Brams and Fishburn, 1988; Brams and Nagel, 1991; Fishburn and Little, 1988). In the case of the IEEE, for example, a close three-way race for president that a petition candidate almost won gave impetus to election reform in that society.[9]

CAV is a somewhat more complex system than approval voting, but it is meant to solve the more difficult problem of ensuring election of candidates representative of two different characteristics, or dimensions, of the electorate.[10] Its major advantages for the professional association were to obviate the need for slate engineering and to give members greater voice in determining the composition of a representative board.

Although CAV was not adopted, experience with both it and approval voting illustrates how rigorous analysis can be used not only to illuminate basic issues of representation but also to find better procedures to ensure PR. In the case of the association, the issue was not just one of achieving fair representation but one of fair division as well: because of the two-dimensional nature of the problem, there had to be a division of seats between both the specialties and the regions such that each category received its fair share.

Doubtless, there are other problems of fair division in elections whose solutions can gain legitimacy from the participation of voters. Implementation of these may well require new election procedures.

[9] In their analysis of the use of approval voting by the IEEE, Brams and Nagel (1991) noted particularly its negative effect on candidates with strong but relatively narrow support.

[10] Other examples of voting systems, related to approval voting for electing sets of candidates are "coalition voting" (Brams and Fishburn, 1992) and "yes–no voting" (Brams and Fishburn, 1993).

11 Conclusions

In this brief concluding chapter, we draw attention to a few general themes that link the more specific findings of the book:

1 Fair division is an old subject with a rich past.

We traced concerns about fair division back to the Bible and ancient Greece, but probably in every culture issues of fair division have arisen and rippled through its institutions. Specific procedures, as well, have a venerable history, which have taken on different forms at different times. For example, divide-and-choose emerged as a legislative procedure in which one house divides and the other chooses in James Harrington's *The Commonwealth of Oceana* in the seventeenth century; nearly 350 years later it is the basis for allocating mining rights under the Convention of the Law of the Sea.

The pervasiveness of a fair-division ethic seems driven in part by its survival value. From a sociobiological perspective, the idea of giving people what they deserve – by reciprocating their help or kindness or, conversely, by not being responsive when they are not – leads to the development of conventions, customs, or even laws that institutionalize the allocation of goods or bads (like chores) with some modicum of fairness. It would be fascinating to study these from the perspective of the theory developed in this book, but we have chosen instead to concentrate on the theoretical analysis of procedures while offering some empirical illustrations.

2 Although there are a plethora of fair-division procedures, they fall into only a few categories.

We have analyzed ten moving-knife procedures and about an equal number of other procedures applicable to both divisible and indivisible goods. This is a dizzying number to keep straight, especially in all their fine detail.

It is precisely for this reason that we have classified the procedures according to the number of players to which they are applicable ($n = 2, 3, 4$, or more) and the properties they satisfy (proportionality, envy-freeness, efficiency, and equitability). We also considered whether the goods or issues being divided were divisible or indivisible, whether or not the players were entitled to equal or unequal shares, and whether or not the division had to be exact or only approximate. The procedures that could be extended to chore division were also identified.

3 Only some of the procedures are applicable to practical fair-division problems.

The list of procedures narrows down quickly when we think about those that actually could be applied to real-life problems. Consequently, the practitioner who must select only one is unlikely to be overwhelmed by a surfeit of possibilities. Many procedures, like Stromquist's three-person envy-free procedure that requires four simultaneously moving knives, will simply not be sensible, despite the fact that this procedure is the only three-person envy-free moving-knife procedure that leads to an allocation that is efficient (in a weak sense).

4 Theory is not antithetical to practice.

Stromquist's procedure is by no means the only procedure that is absurd, practically speaking, to apply to anything except, possibly, a hypothetical cake. Most of the moving-knife procedures fall into this category, and so do many of the discrete procedures, like the Steinhaus–Kuhn lone-divider procedure, which becomes quite complicated even for as few as four players.

But this is not to say that the ideas on which these procedures are based are absurd. For example, several of the procedures guarantee envy-freeness through the creation of ties (e.g., a player stops a moving knife when two pieces are equal, or trims pieces to render them all the same size as a smaller piece). Even if these procedures are not themselves applicable to real-life problems, the idea of tie-making to eliminate envy is a powerful one.

This idea seems to have been invoked on numerous occasions. We briefly recounted the division of Germany and Berlin after World War II into four zones, which was the product of many adjustments in boundaries before an agreement among the allies satisfactory to all was achieved.

Although Germany was united 44 years later in what was, probably, for most Germans and other Europeans a happier solution, a divided Germany from 1945 to 1989 was able to avoid war. On the other hand, war has become all too common in other areas of the world, like the Middle East and the former Yugoslavia, where boundaries are bitterly contested.

Some of the more practical procedures we have proposed and applied, such as adjusted winner (AW) to divorce settlements and the trimming procedure to estate division, probably would not have been developed without the inspiration of the earlier less practical procedures. Theoretical foundations were laid with these earlier procedures, just as we hope some of the ideas we have put forward – such as equitability in the case of AW, and multiple stages in the case of the trimming procedure – might be extended in future work.

5 Practice is not antithetical to theory.

If extensions come, we think they may benefit from experience that practitioners have in trying to apply procedures like AW and the trimming procedure to actual fair-division problems. Is, for example, the fact that the spouses in a divorce each receive the same number of announced points under AW something that pleases them? Or would the spouses prefer a less manipulable procedure, like proportional allocation (PA), even at the price of an inefficient outcome in which both lose points? Or does the combined procedure, whereby AW is used unless one party selects PA as a default option, offer the best of both possible worlds?

It is difficult to answer these questions in the abstract. But experience – or, perhaps better, controlled experiments – with the different point-allocation procedures might help to sort out what properties are most desired by the parties.

These findings, in turn, may help one decide in what direction to pursue further development of the theory. For instance, if the combined procedure is deemed the best choice by participants, we may well ask what are the equilibria that arise under it in a game-theoretic framework, which we have yet to investigate.

6 A stimulus for a good theory is a hard problem.

There is an abundance of problems that motivate the theory of fair division, but those problems that provide the greatest prod tend to be the most intractable. The problem of multiple equilibria in games is a case in point. For example, it is because of the multiplicity of equilibria in divide-the-dollar that the players are prevented from homing in on the 50–50 division, especially if they have greed in their hearts.

The solutions we proposed, which all involved changes in the rules of play and hence the game being played, do not solve the original game but instead place it in a broader class of games. Among what we called reasonable payoff schemes for such games, we found solutions that redounded to the benefit of all the players, enabling them to avoid inefficient outcomes whereby they all end up with little or nothing of the dollar.

Another example of a hard problem is that of finding a constructive and envy-free solution to dividing a cake (or chores) among five or more players. While we have shown that this problem can be solved with the trimming procedure, no moving-knife procedure has yet been found.

There are other unsolved problems related to envy-freeness. For example, the trimming procedure, in its finite incarnation (even for only four players), requires an unbounded number of cuts – that is, the number of cuts depends upon the preferences of the players. By contrast, the three-person Selfridge–Conway discrete procedure requires at most five cuts, regardless of the preferences of the players, and the four-person moving-knife procedure requires at most eleven cuts. Is there an envy-free four-person discrete procedure that requires only a bounded number of cuts?

7 Auction and election procedures should satisfy criteria of fair division.

With auctions, fair division refers to some reasonable allocation of goods, such as to those players who most value them, either when several items are being auctioned off at once, or only one item is being auctioned off but there is recurrent bidding for this kind of item (e.g., building contracts). Major informational problems that beset bidding strategies and lead, among other difficulties, to the winner's curse, are mitigated when auctions are done in two stages, which we argued is fairer not only for the bidders but also probably for the bid taker, who benefits in the long run from better informed bidders who have more stable expectations.

Although everyone agrees that elections should be fair, by fairness we mean something more than honest and free elections. In particular, we suggested as a measure of fairness in the case of legislative elections that parties or other groups should be able to ensure, by themselves, their proportional representation (PR). Our analysis of four well-known PR systems, and one new system (constrained approval voting) that ensured PR of two different characteristics of the electorate, showed that each failed to satisfy certain properties, such as monotonicity, consistency, and nonmanipulability. To select a "best" system, therefore, requires that one decide which desiderata are indispensable and which are not.

8 The study of fair-division procedures can give more structure to theories of equality and justice.

There is a nebulousness in political theory about doctrines such as "all men are created equal" or "equal justice under law." Political philosophers since Aristotle and Plato have offered long discourses on equality and justice, but there is a lack of consensus on the meaning of these protean concepts, as well as related concepts like equity, liberty, and rights.

We do not think our fair-division procedures can settle the debate of whether, for example, justice can be achieved without order, or an order must first be imposed on society – perhaps in violation of individual rights – before justice can be administered. Or, to take another example, must the lot of the worst-off member of society be improved, as Rawls (1971) argues following the maximin criterion, or should help be provided to less needy but possibly more numerous members of society, possibly with a floor constraint for the most needy (Frohlich and Oppenheimer, 1992; Jackson and Hill, 1995). In fact, if there are only two players, AW maximizes the benefit of the worse-off player by equalizing the benefits of both players. Not only does AW ensure equitability, but it also ensures that both players cannot do better (efficiency), which together guarantee envy-freeness.

But there are, as with all procedures, tradeoffs. AW is, at least in principle, highly manipulable. Moreover, if there are more than two players, we know that one cannot guarantee the three properties of equitability, envy-freeness, and efficiency. Although one can always obtain two of these three properties, it is not clear which one to sacrifice – or, indeed, the conditions under which, and the degree to which, any sacrifice will be necessary.

To connect our concepts with the less concrete and more philosophical notions of equality and justice found in the political theory literature will require that bridges be built that link this literature and the literature we have assayed. This will not be easy, especially because the former literature does not usually assume that players start with similar capabilities or endowments. In fact, inequalities of this sort are viewed as a central problem in political theory.

By contrast, both Knaster's procedure and two-stage auctions start from the presumption that players have monetary assets, or can borrow them, with the former procedure also requiring the possibility of side payments to other players. Even when players are assumed not to start with such assets, some of the more complex procedures require that players at least have the intellectual ability to be able to formulate optimal strategies under them. But not all players may have this ability.

Notions of justice that focus on its procedural aspects, such as a person's being entitled to due process or a fair and speedy trial, come closest – at least in spirit – to the notions of fair division analyzed here. To be sure, our focus is much more specific, with its emphasis on guarantees if the players adhere to their prescribed strategies. We believe, nonetheless, that some marriage of our procedures to procedural notions of justice is possible. We encourage an exchange of ideas to help close the gap.

9 Equilibrium results that allow for cooperation and more far-sighted thinking need to be investigated.

We have stressed throughout this book that our procedures enable players to guarantee themselves a certain share by acting strictly on their own, so to speak. These "guarantee" results provide an excellent starting point, but we need to recognize that players may not always desire to play their maximin strategies, which maximize the minimum they receive regardless of the strategies that other players select. Thus, if all players might do better by being more cooperative, then there will be an incentive for the players to try to achieve this more efficient solution.

We need look no further than divide-and-choose to illustrate both the problem and its possible solution. If Bob likes the two ends of the cake and Carol the middle, then giving each his or her preferred piece(s) is an efficient as well as an envy-free allocation. But in the absence of information about each other's preferences, divide-and-choose recommends that the divider split the cake in the middle, making the players envy-free but still leaving them dissatisfied.

Now we showed in chapter 1 that, whoever is the divider, he or she can exploit information about the other's preferences and do better than the aforementioned efficient solution. Although there is a way to make an envy-free and efficient allocation a Nash equilibrium in a game, this game, which assumes common knowledge of preferences, would, practically speaking, be a difficult one for real players to manage. We need more practical procedures to turn envy-free solutions into efficient ones, at least for dividing up heterogeneous goods.

To be sure, we did find a few procedures, like Knaster's and AW, that were efficient and envy-free (and equitable, as well, in the case of AW) in a different context. But most of the *n*-person proportional and envy-free procedures do not ensure efficiency, which in the heterogeneous context is a necessary sacrifice if one wants to maintain an ironclad guarantee of proportionality or envy-freeness.

One might introduce, however, some measure of cooperativeness and shared fate, or some more far-sighted thinking (Brams, 1994a), into the rules of play. Then our present procedures might be rendered less "conservative" in the strategies they prescribe, which in turn could yield all the players a better outcome.

It is a daunting intellectual challenge to deepen and extend the theory, and to redesign practical procedures based on the revised theory, to accomplish this end. We hope, nonetheless, that some readers will be inspired to take up this challenge.

Glossary

This glossary contains definitions, in relatively nontechnical language, of specialized terms used in this book. More rigorous definitions of some concepts can be found in the text, as can descriptions of the more complicated fair-division procedures that do not have simple definitions but whose properties are given here.

Additional-member system. An additional-member system is a proportional-representation (PR) voting system in which some legislators are elected from districts, but new members may be added to the legislature to ensure, insofar as possible, that the parties underrepresented on the basis of their national-vote proportions gain additional seats.

Additivity. A player's preferences are additive if the valuations that he or she attaches to two or more disjoint portions of a good can be summed to obtain his or her total valuation of the combined portions. (In a number of contexts, additivity is equivalent to weak additivity.)

Adjusted district voting (ADV). Adjusted district voting is an additional-member system that results in a variable-size legislature, depending on the number of seats that must be added to satisfy proportional representation (PR).

Adjusted winner (AW). Adjusted winner is a two-person point-allocation fair-division procedure that is envy-free, equitable, and efficient but highly manipulable (in theory if not in practice).

Algorithm. An algorithm is a constructive procedure with finitely many discrete steps.

Approval voting. Approval voting is a voting system in which voters can vote for (i.e., approve of) as many candidates as they like, and the candidate with the most approval votes wins.

Approximate fair-division procedure. A fair-division procedure is approximate if it gives a particular solution that, while not exact, is approximate to any preset tolerance level.

Arbitration. Arbitration of a dispute involves its resolution by an outside party (arbitrator), who makes a decision about the terms of a settlement that is binding on the disputants.

Auction procedure. An auction procedure is one in which players make bids for items, which may be monetary bids or involve points (in which case it is called a point-allocation procedure).

Austin's moving-knife procedure. Austin's moving-knife procedure is a two-person moving-knife procedure in which the players divide a heterogeneous good into two pieces so that both players think it is a 50–50 division. No extension of this procedure to three or more players is known, but it can be extended to yield a division into $k > 2$ pieces that each of two players think is an even division; the resulting division is envy-free but not efficient.

Banach–Knaster last-diminisher procedure. The Banach–Knaster last-diminisher procedure is an n-person discrete fair-division procedure, applicable to a divisible heterogeneous good, that is proportional but neither envy-free nor efficient.

Borda count. The Borda count is a preferential voting system in which points are assigned to candidates, with the lowest-ranked candidate receiving 0 points, the next-lowest 1 point, and so on up, with the candidate receiving the most points winning.

Boundedness. A procedure is bounded if there is a finite upper bound – dependent on the number of players but not their preferences – on the number of cuts that are needed to produce a particular division (e.g., a three-person envy-free division).

C-efficiency. A C-procedure is C-efficient if there is no other C-procedure that yields an allocation that is strictly better for at least one player and as good for all the others.

Cell consistency. Under constrained approval voting, cell consistency occurs when the target election figure (TEF) of a cell that is assigned a larger integer is at least as great as one that is assigned a smaller integer.

Chores problem. The chores problem is the dual of the problem of dividing up goods, in which the object is to give players as few chores (or bads) as possible so that the resulting allocation satisfies certain properties of fairness.

Combined procedure. The combined procedure is a two-person point-allocation procedure in which AW is used at the start – with PA used as a default option if either player requests it – that is envy-free and equitable, whether or not the default option is invoked; it is also efficient if the default option is not used.

Common knowledge. Players in a game have common knowledge when they share certain information, know that they share it, know that they know that they share it, and so on *ad infinitum*.

Common-value auction. A common-value auction is an auction in which the value of an item is the same for all players, but they may have only incomplete information about this value.

Competitiveness assumption (CA). In a two-stage auction, the competitiveness assumption says that players choose preferred strategies (i.e., with higher first-stage bids) by successively eliminating nonpreferred strategies.

Condorcet candidate. A Condorcet candidate is a candidate who can defeat all other candidates, based on majority rule, in separate pairwise contests.

Consistency. Under constrained approval voting (CAV), consistency occurs if there is cell consistency and, in addition, the target election figure (TEF) of every row and column total that is assigned a larger integer is at least as great as one that is assigned a smaller integer.

Constrained approval voting (CAV). Constrained approval voting is a proportional-representation (PR) system in which approval voting is used – subject to constraints placed on the number of candidates that can be elected in different categories – with the candidates most approved of by all the voters, subject to the constraints, winning.

Controlled rounding. A controlled rounding is one in which, for every column and row of a table, the sum of its cell target election figures (TEFs), rounded down or up, equals the column or row (total) TEF, rounded down or up, with the roundings summing to some grand total.

C-procedure. A C-procedure is a discrete or a moving-knife n-person fair-division procedure that results in a set of $n-1$ cuts of a rectangular cake, all of which are parallel to the left and right edges of the cake.

Cumulative voting. Cumulative voting is a proportional-representation (PR) voting system in which each voter can distribute a fixed number of votes among one or more candidates, with the candidates with the most votes winning.

Cyclical majorities. Cyclical majorities occur when majorities of voters prefer candidate x to y, y to z, and z to x, indicating the lack of a social ordering along a single dimension.

Discrete fair-division procedure. A discrete fair-division procedure is one in which players proceed in a step-by-step manner, without relying on the stopping of a continuously moving knife.

Divide-and-choose. Divide-and-choose is a discrete two-person fair-division procedure, applicable to a divisible heterogeneous good, in which one player (the divider) cuts a cake into two pieces, and the other player (the chooser) selects one of the pieces. It is proportional and envy-free but not efficient.

Divide-the-dollar (DD). Divide-the-dollar is a two-person auction procedure in which players make bids for a dollar, which they each receive if the sum of their bids does not exceed a dollar; otherwise, they receive nothing.

DD1, DD2, DD3. These are all variations of DD. DD1 has a reasonable payoff scheme, DD2 involves a second stage, and DD3 has both a reasonable payoff scheme and a second stage.

Divisible good. A divisible good is one that can be divided at any point along a continuum (infinite divisibility), or in discrete units (finite divisibility), without destroying its value.

Dominance inducibility. An outcome is dominance inducible if, given a prior choice by a player, it is dominance solvable.

Dominance solvability. An outcome is dominance solvable if it is the result of the successive elimination of weakly dominated strategies.

Dominant strategy. A dominant strategy is one that dominates (either strongly or weakly) every other strategy.

Dominated strategy. A dominated strategy is one that is dominated (either strongly or weakly) by another strategy.

Domination (strong). A strategy strongly dominates another strategy if – regardless of what strategies the other players choose – this strategy yields a player an outcome that is strictly better than that yielded by the other strategy.

Domination (weak). A strategy weakly dominates another strategy if – regardless of what strategies the other players choose – this strategy yields a player an outcome that is at least as good as that yielded by the other strategy, and, for at least one choice of strategies of the other players, yields a strictly better outcome.

Dubins–Spanier moving-knife procedure. The Dubins–Spanier moving-knife procedure is an *n*-person moving-knife procedure, applicable to a divisible heterogeneous good, that is proportional but neither envy-free nor efficient.

Dutch auction. A Dutch auction is an oral descending auction in which the first person to stop the descent wins.

Efficiency. An allocation is efficient (Pareto-optimal) if there is no other allocation that is strictly better for at least one player and as good for all the others.

Efficiency (auctions). An auction is efficient (Pareto-optimal) if the player with the highest valuation of an item wins it in the auction.

Egalitarian behavior. Egalitarian behavior occurs when all n players in divide-the-dollar (DD), or its variations, choose to bid $1/n$.

Egalitarian outcome. An egalitarian outcome is an allocation in divide-the-dollar (DD), or its variations, that gives to each player the payoff of $100/n$.

Endowment. The endowment of a player is the resources that he or she has at the start of a fair-division procedure.

English auction. An English auction is an oral ascending auction in which the highest bidder wins.

Entitlement. The entitlement of a player specifies the minimal proportion of a good or goods that he or she must receive under a fair-division procedure.

Envy-freeness. An allocation is envy-free if every player thinks he or she receives a portion that is at least tied for largest, or tied for most valuable and, hence, does not envy any other player.

Equitability. An allocation is equitable for two players if each player thinks that the portion he or she receives is worth the same, in terms of his or her valuation, as the portion that the other player receives in terms of that player's valuation. If the two players have different entitlements, equitability means that each player thinks that his or her portion is greater than his or her entitlement by exactly the same percentage.

Even division. A division of a good or goods is even if every player receives, in his or her valuation, exactly the same amount of that good or those goods, making the division proportional and envy-free but not necessarily efficient.

Fairness. A procedure is fair to the degree that it satisfies certain properties (e.g., proportionality, envy-freeness, efficiency, equitability, or invulnerability to manipulation), enabling each player to achieve a certain level of satisfaction with a guarantee strategy.

Filter-and-choose. Filter-and-choose is a procedure in which one player (the filterer) must choose a subset of different goods – by filtering out what he or she considers the best from a larger set – in such a way that the other player (the chooser) will also prefer this portion to the status quo (i.e., no change).

Fink lone-chooser procedure. The Fink lone-chooser procedure is an n-person discrete fair-division procedure, applicable to a divisible heterogeneous good, that is proportional but neither envy-free nor efficient.

Four-person moving-knife procedure. The four-person moving-knife procedure is a four-person procedure, applicable to a heterogeneous good, that is envy-free but not efficient.

Guarantee strategy. A guarantee strategy is a strategy that a player can choose such that no matter what strategies the other players choose, this player can guarantee that his or her portion satisfies certain properties (e.g., is proportional or at least tied for largest or most valuable).

Hamilton allocation. Under constrained approval voting, a Hamilton allocation is one that satisfies the two conditions for the Hamilton method of rounding.

Hare system of single transferable vote (STV). The Hare system of single transferable vote is a preferential voting system that involves the successive elimination of the lowest-vote candidates, and the transfer of surplus votes of those who have already been elected to other candidates.

Heterogeneous good. A heterogeneous good (e.g., a cake with uneven swirls of chocolate and vanilla) is not the same throughout; different players may value a portion of it differently.

Homogeneous good. A homogeneous good is the same throughout; each player values a portion of it the same.

Indivisible good. An indivisible good is one that cannot be divided without destroying its value.

Irrevocable advantage. A player has an irrevocable advantage over another player if the portion of a cake he or she has so far received in a multistage procedure is at least as large as the portion so far received by the other player, together with all of the remaining unallocated cake.

Judicious-bidding assumption (JBA). The judicious-bidding assumption says that in a two-stage auction, a player will never make a stage I bid, or choose a stage II option, such that, if that player wins the auction, he or she might suffer a loss.

Knaster's procedure of sealed bids. Knaster's procedure of sealed bids is an n-person auction procedure, applicable to indivisible homogeneous goods, that is proportional and efficient but not envy-free.

Levmore–Cook moving-knife procedure. The Levmore–Cook moving-knife procedure is a three-person procedure, applicable to a heterogeneous good, that is envy-free but not efficient.

Linearity (under Lucas' method of markers). Linearity implies that n players can create an equal division of all goods into n parts, by placing $n - 1$ markers at points on a straight line, along which the goods to be divided are arrayed from left to right.

Linearity (under point-allocation procedures). Linearity means that the players' marginal utilities are constant – instead of diminishing as one obtains more of something – so, for example, $2x$ percent of a good is twice as good as x percent.

List system. A list system is a proportional-representation (PR) system in which voters can vote for one party, which receives representation in a legislature proportional to the number of votes it receives.

Lucas' method of markers. Lucas' method of markers is an n-person discrete fair-division procedure, applicable to a divisible heterogeneous good or many small indivisible goods, that is proportional but neither envy-free nor efficient.

Manipulability. A procedure is manipulable to the degree that players can, by knowing the preferences of other players, exploit that knowledge to obtain a larger portion of a good or goods than that which they could obtain by not having this knowledge.

Maximin strategy. A maximin strategy maximizes the minimum portion that a player can guarantee for himself or herself, no matter what strategies the other players choose; it is a special case of a guarantee strategy.

Mediation. Mediation of a dispute involves an outside party (mediator), who helps the disputants clarify their objectives, better communicate with each other, find points of agreement, etc., so as to facilitate settlement of the dispute.

Monotonicity. Monotonicity is a property of a preferential voting system indicating that a candidate cannot be hurt by being raised in the preference rankings of some voters while remaining the same in the rankings of all the others.

Moving-knife procedure. A moving-knife procedure is a fair-division procedure that uses a knife, which moves continuously across the surface of a cake, that players can stop and have cuts made.

Nash equilibrium. A Nash equilibrium is a set of strategy choices, associated with an outcome, from which no player would have an incentive to depart unilaterally because his or her departure would lead to a worse, or at least not a better, outcome.

No-benefit constraint. Under adjusted district voting (ADV), the no-benefit constraint prevents the allocation of extra seats to a minority that would result in its having a greater proportion of seats in a legislature than it would have obtained had it won more seats through district elections.

No-regret assumption (NRA). In a two-stage auction, the no-regret assumption says that a player will never make a stage I bid, or choose a stage II option, such that if he or she should turn out to be the player with the highest valuation, he or she might lose the auction.

No-show paradox. The no-show paradox occurs under a preferential voting system when a voter, by not showing up at the polls and voting for a candidate, helps that candidate.

Paradox of voting. A paradox of voting occurs when there are cyclical majorities.

Pareto-optimality. See Efficiency.

Plurality voting. Plurality voting is a voting system in which voters can vote for only one candidate, and the candidate with the most votes wins.

Preferential voting system. A preferential voting system is one in which voters can rank candidates in order of their preference, or in which they can allocate different numbers of votes to them.

Private-value auction. A private-value auction is an auction in which players' valuations are strictly their own and completely unaffected by their knowledge of the other players' valuations or bids.

Point-allocation procedure. A point-allocation procedure is a procedure under which players can allocate a fixed number of points (say, 100) to different goods (or issues) that reflects, if they are truthful, the importance they attach to receiving these goods (or winning on these issues).

Procedure (fair division). A fair-division procedure is a process for allocating a good or goods that proceeds in a step-by-step fashion. A step may involve the use of devices like moving knives, whose use is governed by the rules.

Proportional allocation (PA). Proportional allocation is a two-person point-allocation procedure that is envy-free, equitable, and relatively non-manipulable, but it is not efficient.

Proportionality. An allocation is proportional if every one of n players thinks he or she received a portion that has a size or value of at least $1/n$.

Proportionality constraint. Under adjusted district voting (ADV), the proportionality constraint prevents the seats of a minority in the legislature from exceeding its proportion of votes in the electorate.

Proportional representation (PR) system. A proportional-representation system is one in which the seats in a legislature are apportioned among groups according to the amount of their support in the electorate.

Public good. A public good is a good from which each player can benefit without detracting from the benefit of other players. (When the benefit is a loss, the public good is a *public bad.*)

Quota (constrained approval voting [CAV]). Quota is satisfied when the apportionment of seats to a cell is the target election figure (TEF), rounded either down or up.

Quota (Hare system of single transferable vote [STV]). The quota is the smallest number of votes that a candidate must receive to ensure his or her election.

Reasonable payoff scheme. A reasonable payoff scheme in divide-the-dollar (DD) and its variations is one that satisfies five conditions that do not entail harsh punishment.

Rectifiability. A procedure is rectifiable if it enables a player to "correct" a first-stage choice in a second stage.

Referee. A referee is a person who ensures that the rules of a procedure are followed.

Rules. Rules describe the choices that players can make at each stage of a procedure, which can be enforced by a referee who does not know their preferences.

Selfridge–Conway discrete procedure. The Selfridge–Conway discrete procedure is a three-person fair-division procedure, applicable to a heterogeneous good, that is bounded and envy-free but not efficient.

Sensitivity analysis. Sensitivity analysis involves determining how sensitive outcomes are to small changes in player strategies.

Separability. Goods (or issues) are separable if the values players attach to them are independent of the players' possessing other goods (or winning on other issues).

Slate engineering. Slate engineering is the rigging of an election through the selection of a particular slate of candidates designed to help some candidates and hurt others.

Solution (fair division). A solution to a fair-division problem is a procedure, together with a description of a strategy on how to apply its rules, that – if followed – guarantees a certain level of satisfaction (e.g., an allocation that is envy-free).

Split award. A split award in an auction occurs when a divisible good, instead of going to a single winner, is apportioned among two or more winning bidders.

Steinhaus–Kuhn lone-divider procedure. The Steinhaus–Kuhn lone-divider procedure is an n-person discrete fair-division procedure, applicable to a divisible heterogeneous good, that is proportional but neither envy-free nor efficient.

Strategy (in fair division). A strategy is advice to a player about the choices he or she *should* make, based on that player's preferences, that are consistent with a procedure's rules and give a solution having certain properties.

Strategy (in game theory). A strategy is a complete plan that specifies the course of action a player will follow, depending on the strategies of the other players.

Stromquist moving-knife procedure. The Stromquist moving-knife procedure is a three-person fair-division procedure, applicable to a heterogeneous good, that is envy-free and C-efficient but not efficient.

Target election figure (TEF). Under constrained approval voting (CAV), a target election figure is the exact number of seats to which each cell in a table – which breaks down the electorate into different categories (e.g., by region and specialty) – is entitled.

Trimming procedure. The trimming procedure is a discrete n-person fair-division procedure, applicable to a heterogeneous good or several homogeneous goods, that is envy-free but not efficient; there is an infinite version and a more complex finite version, which is unbounded.

Two-stage auction. A two-stage auction is an auction in which players submit sealed bids in stage I, which are then made public; in stage II, a player may affirm his or her own bid or usurp another player's bid; the highest stage II bid wins (unless there is a tie, in which case the player who bid highest in stage I breaks the tie and wins).

Utility. The utility of some portion of a good or goods to a player is a numerical value indicating that player's degree of preference for that portion; if the player prefers one portion to another, then the former portion receives a higher numerical value.

Valuation. A player's valuation – as opposed to his or her announced valuation – is the utility he or she attaches to obtaining a good (or winning on an issue).

Vickrey (second-price) auction. A Vickrey (second-price) auction is a sealed-bid auction in which the highest bidder wins but pays only the second-highest bid.

Weak additivity. A player's preferences are weakly additive if, whenever he or she prefers A to B and C to D, and there is no overlap between A and C, that player prefers A together with C to B together with D.

Webb moving-knife procedure. The Webb moving-knife procedure is a three-person fair-division procedure, applicable to a heterogeneous good, that is envy-free but not efficient.

Winner's curse. A winner's curse in an auction occurs when the winner, by virtue of winning, ends up overpaying and, in this sense, "losing."

Bibliography

Abreu, Dilip and H. Matsushima (1994). "Exact Implementation." *Journal of Economic Theory* 64 (1, October): 1–19.
Akin, Ethan (1995). "Wilfredo Pareto Cuts the Cake." *Journal of Mathematical Economics* 24 (1): 23–44.
Alkan, Ahmet, Gabrielle Demange, and David Gale (1991). "Fair Allocation of Indivisible Goods and Criteria of Justice." *Econometrica* 59 (4, July): 1023–39.
Alon, Noga (1987). "Splitting Necklaces." *Advances in Mathematics* 63 (3, March): 247–53.
Andrews, Edmund L. (1995). "Winners of Wireless Auction to Pay $7 billion." *New York Times*, March 14: D1, D3.
Anton, James and Dennis A. Yao (1989). "Split Awards, Procurement, and Innovation." *RAND Journal of Economics* 20 (4, Winter): 538–52.
(1992). "Coordination in Split Award Auctions." *Quarterly Journal of Economics* 107 (2, May): 681–707.
Aragones, Enriqueta (1995). "A Derivation of the Money Rawlsian Solution." *Social Choice and Welfare*, forthcoming.
Arnsperger, Christian (1994). "Envy-Freeness and Distributive Justice." *Journal of Economic Surveys* 8 (2, June): 155–86.
Arrow, Kenneth J. (1951). *Social Choice and Individual Values* (2nd edn., 1963). New Haven, CT: Yale University Press.
(1992). "I Know a Hawk from a Handsaw." In Michael Szenberg (ed.), *Eminent Economists: Their Life Philosophies*. Cambridge: Cambridge University Press, pp. 42–50.
Aumann, Robert J. and Michael Maschler (1985). "Game Theoretic Analysis of a Bankruptcy Problem from the Talmud." *Journal of Economic Theory* 36 (2, August): 195–213.
Austin, A. K. (1982). "Sharing a Cake." *Mathematical Gazette* 66 (437, October): 212–15.
Austin, A. K. and Walter Stromquist (1983). "Commentary." *American Mathematical Monthly* 90 (7, August–September): 474.
Bacharach, Michael (1993). "Variable Universe Games." In Ken Binmore, Alan Kirman, and Piero Tani (eds.), *Frontiers of Game Theory*. Cambridge, MA: MIT Press, pp. 255–75.

Balinski, Michel L. and Gabrielle Demange (1989). "An Axiomatic Approach to Proportionality between Matrices." *Mathematics of Operations Research* 14 (4, November): 700–190.

Balinski, Michel L. and H. Peyton Young (1982). *Fair Representation: Meeting the Ideal of One Man, One Vote*. New Haven, CT: Yale University Press.

Barbanel, Julius B. (1995a). "Game-Theoretic Algorithms for Fair and Strongly Fair Cake Division with Entitlements." *Colloquin Mathematicum* 69: 59–73.

(1995b). "On the Possibilities of Partitioning a Cake." *Proceedings of the American Journal of Mathematical Society* 123 (7, July): 2061–70.

(1995c). "Super Envy-Free Cake Division and Independence of Measures." *Mathematical Analysis and Applications*, forthcoming.

Barbanel, Julius B. and Alan D. Taylor (1995). "Preference Relations and Measures in the Context of Fair Division." *Proceedings of the American Mathematical Society*, forthcoming.

Barbanel, Julius B. and William S. Zwicker (1994). "Two Applications of a Theorem of Dvoretsky, Wald, and Wolfovitz to Cake Division." Preprint, Department of Mathematics, Union College.

Baumol, William J. (1986). *Superfairness: Applications and Theory*. Cambridge, MA: MIT Press.

Beck, Anatole (1987). "Constructing a Fair Border." *American Mathematical Monthly* 94 (2, February): 157–62.

Bennett, Sandi *et al.* (1987). "Fair Divisions: Getting Your Share." HIMAP [High School Mathematics and Its Applications] Module 9.

Ben Ze'ev, Aaron (1992). "Envy and Inequality." *Journal of Philosophy* 89 (11, November): 551–81.

Berliant, Marcus, Karl Dunz, and William Thomson (1992). "On the Fair Division of a Heterogeneous Commodity." *Journal of Mathematical Economics* 21 (3): 201–16.

Bernheim, B. Douglas (1984). "Rationalizable Strategic Behavior." *Econometrica* 52 (4, July): 1007–28.

Binmore, Ken (1992). *Fun and Games: A Text on Game Theory*. Lexington, MA: D. C. Heath.

(1994). *Game Theory and the Social Contract. Volume I: Playing Fair*. Cambridge, MA: MIT Press.

Black, Duncan (1958). *The Theory of Committees and Elections*. Cambridge: Cambridge University Press.

Blitzer, Charles (1960). *An Immortal Commonwealth: The Political Thought of James Harrington*. New Haven, CT: Yale University Press.

Boggs, S. Whittemore (1940). *International Boundaries: A Study of Boundary Functions and Problems*. New York: Columbia University Press.

Bolger, Edward M. (1985). "Monotonicity and Other Paradoxes in Some Proportional Representation Schemes." *SIAM Journal on Algebraic and Discrete Methods* 6 (2, April): 283–91.

Bossert, Walter (1993). "An Alternative Solution to Bargaining Problems with Claims." *Mathematical Social Sciences* 25 (3, May): 205–20.

Brams, Steven J. (1975). *Game Theory and Politics*. New York: Free Press.

(1980). *Biblical Games: A Strategic Analysis of Stories in the Old Testament*. Cambridge, MA: MIT Press.

(1982). "The AMS Nomination Procedure is Vulnerable to 'Truncation of Preferences.'" *Notices of the American Mathematical Society* 29 (2, February): 136–8.

(1985a). *Rational Politics: Decisions, Games, and Strategy*. Washington, DC: CQ Press (reprinted by Academic Press, 1989).

(1985b). *Superpower Games: Applying Game Theory to Superpower Conflict*. New Haven, CT: Yale University Press.

(1988). "MAA Elections Produce Decisive Winners." *Focus: The Newsletter of the Mathematical Association of America* 8 (3, May–June): 1–2.

(1990a). "Constrained Approval Voting: A Voting System to Elect a Governing Board." *Interfaces* 20 (5, September–October): 67–79.

(1990b). *Negotiation Games: Applying Game Theory to Bargaining and Arbitration*. New York: Routledge.

(1994a). *Theory of Moves*. Cambridge: Cambridge University Press.

(1994b). "Voting Procedures." In Robert J. Aumann and Sergiu Hart (eds.), *Handbook of Game Theory with Economic Applications*, vol. II. Amsterdam: Elsevier Science, pp. 1055–89.

Brams, Steven J. and Peter C. Fishburn (1983). *Approval Voting*. Cambridge, MA: Birkhäuser Boston.

(1984a). "A Note on Variable-Size Legislatures to Achieve Proportional Representation." In Arend Lijphart and Bernard Grofman (eds.), *Choosing an Electoral System: Issues and Alternatives*. New York: Praeger, pp. 175–7.

(1984b). "Proportional Representation in Variable-Size Legislatures." *Social Choice and Welfare* 1 (3, October): 211–29.

(1984c). "Some Logical Defects in the Single Transferable Vote." In Arend Lijphart and Bernard Grofman (eds.), *Choosing an Electoral System: Issues and Alternatives*. New York: Praeger, pp. 147–51.

(1988). "Does Approval Voting Elect the Lowest Common Denominator?" *PS: Political Science and Politics* 21 (2, Spring): 277–84.

(1991). "Alternative Voting Systems." In L. Sandy Maisel (ed.), *Political Parties & Elections in the United States: An Encyclopedia*, vol. I. New York: Garland, pp. 23–31.

(1992a). "Approval Voting in Scientific and Engineering Societies." *Group Decision and Negotiation* 1 (1, April): 41–55.

(1992b). "Coalition Voting." *Mathematical and Computer Modelling (Formal Theories of Politics II: Mathematical Modelling in Political Science)* 16 (8/9, August/September): 15–26.

(1993). "Yes–No Voting." *Social Choice and Welfare* 10 (1): 35–50.

Brams, Steven J. and D. Marc Kilgour (1988). *Game Theory and National Security*. New York: Basil Blackwell.

(1995). "Bargaining Procedures That Induce Honesty." *Group Decision and Negotiation*, forthcoming.

Brams, Steven J., D. Marc Kilgour, and Samuel Merrill, III (1991). "Arbitration Procedures." In H. Peyton Young (ed.), *Negotiation Analysis*. Ann Arbor, MI: University of Michigan Press, pp. 47–65.

Brams, Steven J. and Jack H. Nagel (1991). "Approval Voting in Practice." *Public Choice* 71 (1–2, August): 1–17.

Brams, Steven J., and Philip D. Straffin, Jr. (1982). "The Apportionment Problem." *Science* 217 (4558, July 30): 437–8.

Brams, Steven J. and Alan D. Taylor (1994a). "Divide the Dollar: Three Solutions and Extensions." *Theory and Decision* 37 (2, September): 211–31.

(1994b). "Fair Division by Point Allocation." Preprint, Department of Politics, New York University.

(1994c). "A Procedure for Divorce Settlement." Preprint, Department of Politics, New York University.

(1994d). "Two-Stage Auctions I: Private-Value Strategies." Preprint, Department of Politics, New York University.

(1994e). "Two-Stage Auctions II: Common-Value Strategies and the Winner's Curse." Preprint, Department of Politics, New York University.

(1995a). "An Envy-Free Cake Division Protocol." *American Mathematical Monthly* 102 (1, January): 9–18.

(1995b). "A Note on Envy-Free Cake Division." *Journal of Combinatorial Theory*, Series A 70 (1, April): 170–3.

Brams, Steven J., Alan D. Taylor, and William S. Zwicker (1995a). "A Moving-Knife Solution to the Four-Person Envy-Free Cake Division Problem." *Proceedings of the American Mathematical Society*, forthcoming.

(1995b). "Old and New Moving-Knife Schemes." *Mathematical Intelligencer* 17 (4), forthcoming.

Broome, John (1991). *Weighing Goods*. Oxford, UK: Basil Blackwell.

Campbell, Paul J. (1994). "Mathematics." *1995 Encyclopaedia Britannica Yearbook of Science and the Future*. Chicago: Encyclopaedia Britannica, pp. 379–83.

Capen, E., R. Clapp, and W. Campbell (1971). "Competitive Bidding in High-Risk Situations." *Journal of Petroleum Technology* 23 (June): 641–53.

Chamberlin, John R. and Paul N. Courant (1983). "Representative Deliberations and Representative Decisions: Proportional Representation and the Borda Rule." *American Political Science Review* 77 (3, September): 718–33.

Chatterjee, K. and L. Samuelson (1990). "Perfect Equilibria in Simultaneous-Offers Bargaining." *International Journal of Game Theory* 19 (3): 237–67.

Chisholm, Michael and David M. Smith (eds.) (1990). *Shared Space: Divided Space: Essays on Conflict and Territorial Organization*. London: Unwin Hyman.

Chun, Youngsub (1988). "The Proportional Solution for Rights Problems." *Mathematical Social Sciences* 15 (3, June): 231–46.

Chun, Youngsub and William Thomson (1992). "Bargaining Problems with Claims." *Mathematical Social Sciences* 24 (1, August): 19–33.

Cohen, I. Bernard (1994). *Interactions: Some Contacts between the Natural Sciences and the Social Sciences*. Cambridge, MA: MIT Press.

COMAP [Consortium for Mathematics and Its Applications] (1994). *For All Practical*

Purposes: Introduction to Contemporary Mathematics, 3rd edn. New York: W. H. Freeman.

Conway, John H. (1993). Private communication to A. D. Taylor (May 17).

Cox, Lawrence H. (1987). "A Constructive Procedure for Unbiased Controlled Rounding." *Journal of the American Statistical Association* 82 (398, June): 520–4.

Cox, Lawrence H. and Lawrence R. Ernst (1982). "Controlled Rounding." *INFOR* 20 (4, November): 423–32.

Crawford, Vincent (1977). "A Game of Fair Division." *Review of Economic Studies* 44 (2, June): 235–47.

(1979). "A Procedure for Generating Pareto-Efficient Egalitarian-Equivalent Allocations." *Econometrica* 47 (1, January): 49–60.

Curiel, Imma J., Michael Maschler, and Stef H. Tijs (1988). "Bankruptcy Games." *Zeitschrift für Operations Research* 31 (1): A143–59.

Custer, Carolyn E. (1994). "Cake-Cutting Hugo Steinhaus Style: Beyond N = 3." Senior Thesis, Department of Mathematics, Union College.

Dagan, Nir and Oscar Volij (1993). "The Bankruptcy Problem: A Cooperative Bargaining Approach." *Mathematical Social Sciences* 26 (3, November): 287–97.

Davis, Robert L. (ed.) (1955). *Elementary Mathematics of Sets with Applications.* Department of Mathematics, Tulane University (reprinted by Mathematical Association of America, Washington, DC, 1958).

de la Mora, Gonzalo Fernández (1987). *Egalitarian Envy: The Political Foundations of Social Justice.* New York: Paragon House.

Demange, Gabrielle (1984). "Implementing Efficient Egalitarian Equivalent Allocations." *Econometrica* 52 (5, September): 1167–77.

Demko, Stephen and Theodore P. Hill (1988). "Equitable Distribution of Indivisible Objects." *Mathematical Social Sciences* 16 (2, October): 145–58.

Doron, Gideon and Richard Kronick (1977). "Single Transferable Vote: An Example of a Perverse Social Choice Function." *American Journal of Political Science* 21 (2, May): 303–11.

Downs, Michael (1977). *James Harrington.* Boston: G. K. Hall.

Dubins, Lester E. (1977). "Group Decision Devices." *American Mathematical Monthly* 84 (5, May): 350–6.

Dubins, Lester E. and E. H. Spanier (1961). "How to Cut a Cake Fairly." *American Mathematical Monthly* 68 (1, January): 1–17.

Dummett, Michael (1984). *Voting Procedures.* Oxford: Oxford University Press.

Ehtamo, Harri, Marko Verkama, and Raimo P. Hämäläinen (1994). "Negotiating Efficient Agreements over Continuous Issues." Preprint, Systems Analysis Laboratory, Helsinki University of Technology (June).

Elster, Jon (1989). *Solomonic Judgments: Studies in the Limitations of Rationality.* Cambridge: Cambridge University Press.

(1992). *Local Justice: How Institutions Allocate Scarce Goods and Necessary Burdens.* New York: Russell Sage Foundation.

Elster, Jon and John E. Roemer (eds.) (1991). *Interpersonal Comparisons of Well-Being.* Cambridge: Cambridge University Press.

Elton, John, Theodore P. Hill, and Robert P. Kertz (1986). "Optimal-Partitioning

Inequalities for Nonatomic Probability Measures." *Transactions of the American Mathematical Society* 296 (2, August): 703–25.

Engelbrecht-Wiggans, Richard (1988). "On a Possible Benefit to Bid Takers from Using Multi-Stage Auctions." *Management Science* 34 (9, September): 1109–20.

Erlanger, Howard S., Elizabeth Chambliss, and Marygold S. Melli (1987). "Participation and Flexibility in Informal Processes: Cautions from the Divorce Context." *Law and Society Review* 21 (4): 585–604.

Ernst, Lawrence R. (1994). "Apportionment Methods of the House of Representatives and the Court Challenges." *Management Science* 40 (10, October): 1207–27.

Even, S. and A. Paz (1984). "A Note on Cake Cutting." *Discrete Applied Mathematics* 7: 285–96.

Fagan, J. T., B. V. Greenberg, and B. Hemmig (1988). "Controlled Rounding of Three Dimensional Tables." Statistical Research Division Report Series, Census/SRD/RR-88/02, Bureau of Census, US Department of Commerce.

Feldman, Allan and Alan Kirman (1974). "Fairness and Envy." *American Economic Review* 64 (6, December): 995–1005.

Fineman, Martha Albertson (1991). *The Illusion of Equality: The Rhetoric and Reality of Divorce Reform*. Chicago: University of Chicago Press.

Fink, A. M. (1964). "A Note on the Fair Division Problem." *Mathematics Magazine* 37 (5, November–December): 341–2.

(1994). Private communication to S. J. Brams and A. D. Taylor (December 1).

(1995). Private communication to S. J. Brams (February 22).

Finnie, David H. (1992). *Shifting Lines in the Sand: Kuwait's Elusive Frontier with Iraq*. Cambridge, MA: Harvard University Press.

Fishburn, Peter C. (1974a). "Lexicographic Orders, Utilities and Decision Rules: A Survey." *Management Science* 20 (11, July): 1442–71.

(1974b). "Paradoxes of Voting." *American Political Science Review* 68 (2, June): 537–46.

(1982). "Monotonicity Paradoxes in the Theory of Elections." *Discrete Applied Mathematics* 4 (2, April): 119–34.

Fishburn, Peter C. and Steven J. Brams (1983). "Paradoxes of Preferential Voting." *Mathematics Magazine* 56 (4, September): 207–14.

(1984). "Manipulability of Voting by Sincere Truncation of Preferences." *Public Choice* 44 (3): 397–410.

Fishburn, Peter C. and John D. C. Little (1988). "An Experiment in Approval Voting." *Management Science* 34 (5, May): 555–68.

Fishburn, Peter C. and Rakesh K. Sarin (1994a). "Fairness and Social Risk I: Unaggregated Analyses." *Management Science* 40 (9, September): 1174–88.

(1994b). "Fairness and Social Risk II: Aggregated Analyses." Preprint, AT&T Bell Laboratories, Murray Hill, NJ.

Fisher, Ronald A. (1938). "Quelques Remarques Sur L'Estimation en Statistique." *Biotypologie*: 153–9.

Fleurbaey, Marc (1992). "L'absence d'envie: un critère de justice?" Preprint, Départment Recherche, INSEE, Malakoff, France.

(1994). "On Fair Compensation." *Theory and Decision* 36 (3, May): 277–307.

Foley, Duncan (1967). "Resource Allocation and the Public Sector." *Yale Economic Essays* 7 (1, Spring): 45–98.

Forsythe, Robert *et al.* (1994). "Fairness in Simple Bargaining Experiments." *Games and Economic Behavior* 6 (3, May): 347–69.

French, Simon (1986). *Decision Theory: An Introduction to the Mathematics of Rationality*. New York: Halsted/John Wiley & Sons.

Frohlich, Norman and Joe Oppenheimer (1992). *Choosing Justice: An Experimental Approach to Ethical Theory*. Berkeley, CA: University of California Press.

Fudenberg, Drew and Jean Tirole (1991). *Game Theory*. Cambridge, MA: MIT Press.

Gale, David (1993). "Mathematical Entertainments." *Mathematical Intelligencer* 15 (1, Winter): 48–52.

Gale, John, Kenneth G. Binmore, and Larry Samuelson (1995). "Learning to be Imperfect: The Ultimatum Game." *Games and Economic Behavior* 8 (1, January): 56–90.

Gamow, George and Marvin Stern (1958). *Puzzle-Math*. New York: Viking.

Gardner, Lloyd C. (1993). *Spheres of Influence: The Great Powers Partition Europe, from Munich to Yalta*. Chicago: Ivan R. Dee.

Gardner, Martin (1978). *Aha! Insight*. New York: W. F. Freeman.

Gassner, Marjorie B. (1991). "Biproportional Delegations: A Solution for Two-Dimensional Proportional Representation." *Journal of Theoretical Politics* 3 (3, July): 321–42.

Gilbert, Margaret (1989). "Rationality and Salience." *Philosophical Studies* 57 (1, September): 6l–77.

(1990). "Rationality, Coordination, and Convention." *Synthese* 84 (1, July): 1–21.

Gilpin, Kenneth N. (1991). "3-Year Note Auction Goes Badly." *New York Times* (November 6), D1, D15.

Glazer, Jacob and Ching-to Albert Ma (1989). "Efficient Allocation of a 'Prize' – King Solomon's Dilemma." *Games and Economic Behavior* 1 (3): 223–33.

Gleick, Peter H. (1993). "Water and Conflict: Fresh Water Resources and International Security." *International Security* 18 (1, Summer): 79–112.

Goertz, Gary and Paul F. Diehl (1992). *Territorial Changes and International Conflict*. London: Routledge.

Goodwin, Barbara (1992). *Justice by Lottery*. Chicago: University of Chicago Press.

Gulick, Edward Vose (1955). *Europe's Classical Balance of Power: A Case History of the Theory and Practice of One of the Great Concepts of European Statecraft*. New York: W. W. Norton.

Gusfield, Dan and Robert W. Irving (1989). *The Stable Marriage Problem: Structure and Algorithms*. Cambridge, MA: MIT Press.

Güth, Werner and Reinhard Tietz (1990). "Ultimatum Bargaining Behavior: A Survey and Comparison of Experimental Results." *Journal of Economic Psychology* 11 (3, September): 417–49.

Hardin, Russell (1982). *Collective Action*. Baltimore: Johns Hopkins University Press.

Harrington, James (1656). *The Commonwealth of Oceana* and *A System of Politics*, ed. J. G. A. Pocock. Cambridge: Cambridge University Press, 1992.

Harstad, Ronald M. (1990). "Alternative Common-Value Auction Procedures: Revenue Comparisons with Free Entry." *Journal of Political Economy* 98 (2, April): 421–9.

Harstad, Ronald M. and Michael H. Rothkopf (1995). "Withdrawable Bids as Winner's Curse Insurance." *Operations Research*, forthcoming.

Hill, Theodore P. (1983). "Determining a Fair Border." *American Mathematical Monthly* 90 (7, August–September): 438–42.

(1987a). "Partitioning General Probability Measures." *Annals of Probability* 15 (2, June): 804–13.

(1987b). "A Sharp Partitioning-Inequality for Non-Atomic Probability Measures Based on the Mass of the Infimum of the Measures." *Probability Theory and Related Fields* 75 (1, May): 143–7.

(1993). "Stochastic Inequalities." *IMS Lecture Notes* 22: 116–32.

Hively, Will (1995). "Dividing the Spoils." *Discover* 16 (3, March): 49–57.

Hobby, Charles R. and John R. Rice (1965). "A Moment Problem in L_1 Approximation." *Proceedings of the American Mathematical Society* 16: 665–70.

Holcombe, Randall G. (1983), "Applied Fairness Theory: Comment." *American Economic Review* 73 (5, December): 1153–6.

Holzman, Ron (1988/1989). "To Vote or Not to Vote: What Is the Quota?" *Discrete Applied Mathematics* 22 (2, April): 133–41.

Hopmann, P. Terrence (1991). "'I Cut, You Choose': A Model for Negotiating Tradeoffs in Complex, Multi-Issue Negotiations." Preprint, Department of Political Science, Brown University.

Hylland, Aanund and Richard Zeckhauser (1979). "The Efficient Allocation of Individuals to Positions." *Journal of Political Economy* 87 (2, April): 293–314.

Jackson, Michael and Peter Hill (1995). "A Fair Share." *Journal of Theoretical Politics* 7 (2, April): 169–79.

Jackson, Michael O., Thomas R. Palfrey, and Sanjay Srivastava (1994). "Undominated Nash Implementation in Bounded Mechanisms." *Games and Economic Behavior* 6 (3, May): 474–501.

Kagel, John H. (1995). "Auctions." In Alvin E. Roth and John H. Kagel (eds.), *Handbook of Experimental Economics*. Princeton, NJ: Princeton University Press.

Kalai, Ehud and Meir Smorodinsky (1975). "Other Solutions to Nash's Bargaining Problem." *Econometrica* 43 (3, May): 513–18.

Keefer, Donald L., F. Beckley Smith, Jr., and Harry B. Back (1991). "Development and Use of a Modelling System to Aid a Major Oil Company in Allocating Bidding Capital." *Operations Research* 39 (1, January–February): 28–41.

Keeney, Ralph L. (1992). *Value-Focused Thinking: A Path to Creative Decision-making*. Cambridge, MA: Harvard University Press.

Keeney, Ralph L. and Howard Raiffa (1991). "Structuring and Analyzing Values for Multiple-Issue Negotiations." In H. Peyton Young (ed.), *Negotiation Analysis*. Ann Arbor, MI: University of Michigan Press, pp. 131–51.

Kelly, James P., Bruce L. Golden, and Arjang A. Assad (1989). "The Controlled Rounding Problem: A Review." Working Paper Series MS/S 89-016, College of Business and Management, University of Maryland (July).

Kelly, Jerry S. (1987). *Social Choice Theory: An Introduction*. New York: Springer-Verlag.

Kim, K. H., F. W. Roush, and M. D. Intriligator (1992). "Overview of Mathematical Social Sciences." *American Mathematical Monthly* 99 (9, November): 838–44.

Kleinman, Danny (1994). Private communication to S. J. Brams (September 20).

Klotz, Dorothy E. and Kalyan Chatterjee (1995). "Variable Split Awards in a Single-Stage Procurement Model." *Group Decision and Negotiation* 4 (4, July): 295–310.

Knaster, B. (1946). "Sur le Problème du Partage Pragmatique de H. Steinhaus." *Annales de la Societé Polonaise de Mathematique* 19: 228–30.

Kolm, Serge-Christophe (1995). *Modern Theories of Justice*. Cambridge, MA: MIT Press.

Kressel, Kenneth (1985). *The Process of Divorce: How Professionals and Couples Negotiate Settlements*. New York: Basic.

Kuhn, Harold W. (1967). "On Games of Fair Division." In Martin Shubik (ed.), *Essays in Mathematical Economics in Honor of Oskar Morgenstern*. Princeton, NJ: Princeton University Press, pp. 29–37.

Lax, David A. and James K. Sebenius (1986). *The Manager as Negotiator: Bargaining for Cooperation and Competitive Gain*. New York: Free Press.

LeGrand, Julian (1991). *Equity and Choice: An Essay in Economics and Applied Philosophy*. London: Harper Collins Academic.

Legut, Jerzy (1985). "The Problem of Fair Division for Countably Many Participants." *Journal of Mathematical Analysis and Applications* 109 (1, July): 83–9.

(1986). "Market Games with a Continuum of Indivisible Commodities." *International Journal of Game Theory* 15 (1): 1–7.

(1987). "A Game of Fair Division with Continuum of Players." *Colloquium Mathematicum* 53 (fasc. 2): 323–31.

(1988a). "A Game of Fair Division in the Normal Form." *Colloquium Mathematicum* 56 (fasc. 1): 179–84.

(1988b). "Inequalities for α-Optimal Partitioning of a Measurable Space." *Proceedings of the American Mathematical Society* 104 (4, December): 1249–51.

(1990). "On Totally Balanced Games Arising from Cooperation in Fair Division." *Games and Economic Behavior* 2 (1, March): 47–60.

Legut, Jerzy, J. A. M. Potters, and S. H. Tijs (1994). "Economies with Land – A Game-Theoretical Approach." *Games and Economic Behavior* 6 (3, May): 416–30.

Legut, Jerzy and Maciej Wilczýnski (1988). "Optimal Partitioning of a Measurable Space." *Proceedings of the American Mathematical Society* 104 (1, September): 262–4.

Leng, Russell J. and William Epstein (1985). "Calculating Weapons Reductions." *Bulletin of the Atomic Scientists* 41 (2, February): 39–41.

Levmore, Saul X. and Elizabeth Early Cook (1981). *Super Strategies for Puzzles and Games*. Garden City, NY: Doubleday.

Lewis, David (1969). *Convention: A Philosophic Study*. Cambridge, MA: Harvard University Press.

Lowi, Miriam R. (1993a). "Bridging the Divide: Transboundary Resource Disputes and the Case of West Bank Water." *International Security* 18 (1, Summer): 113–38.

(1993b). *Water and Power: The Politics of a Scarce Resource in the Jordan River Basin*. Cambridge: Cambridge University Press.

Lowry, S. Todd (1987). *The Archeology of Economic Ideas: The Classical Greek Tradition*. Durham, NC: Duke University Press.

Lucas, William F. (1994). Private communication to S. J. Brams (October 19).

Luce, R. Duncan and Howard Raiffa (1957). *Games and Decisions: Introduction and Critical Survey*. New York: Wiley.

Ludwin, William G. (1978). "Strategic Voting and the Borda Method." *Public Choice* 33 (1): 85–90.

Lustick, Ian S. (1993). *Unsettled States, Disputed Lands: Britain and Ireland, France and Algeria, Israel and the West Bank-Gaza*. Ithaca, NY: Cornell University Press.

Lyapounov, A. (1940). "Sur les Fonctions-Vecteurs Complétement Additives." *Bulletin of the Academy of the Science USSR* 4: 465–78.

McAfee, R. Preston and John McMillan (1987). "Auctions and Bidding." *Journal of Economic Literature* 25 (2, June): 699–738.

McAvaney, Kevin, Jack Robertson, and William Webb (n.d.). "Ramsey Partitions of Integers and Fair Divisions." Preprint, Department of Mathematics, Washington State University.

Maccoby, Eleanor E. and Robert H. Mnookin (1992). *Dividing the Child: Social and Legal Dilemmas of Custody*. Cambridge, MA: Harvard University Press.

Mailath, George J. and Peter Zemsky (1991). "Collusion in Second Price Auctions with Heterogeneous Bidders." *Games and Economic Behavior* 3 (4, November): 467–86.

Maskin, Eric S. (1987). "On the Fair Allocation of Indivisible Goods." In George R. Feiwel (ed.), *Arrow and the Foundations of the Theory of Economic Policy*. New York: New York University Press, pp. 341–9.

Maurer, Stephen B. (1985). "The Algorithmic Way of Life Is Best." *College Mathematics Journal* 16 (1, January): 2–5.

Merrill, Samuel, III (1988). *Making Multicandidate Elections More Democratic*. Princeton, NJ: Princeton University Press.

Milgrom, Paul (1989). "Auctions and Bidding: A Primer." *Journal of Economic Perspectives* 3 (3, Summer): 3–22.

Mill, John Stuart (1862). *Considerations on Representative Government*. New York: Harper and Brothers.

Miner, Steven Merritt (1988). *Between Churchill and Stalin: The Soviet Union, Great Britain, and the Origins of the Grand Alliance*. Chapel Hill, NC: University of North Carolina Press.

Moulin, Hervé (1979). "Dominance Solvable Voting Schemes." *Econometrica* 47 (6, November): 1337–51.

(1984). "Implementing the Kalai-Smorodinsky Bargaining Solution." *Journal of Economic Theory* 33 (1, June): 32–45.

(1988a). *Axioms of Cooperative Decision Making*. Cambridge: Cambridge University Press.

(1988b). "Condorcet's Principle Implies the No Show Paradox." *Journal of Economic Theory* 45 (1, June): 53–64.

(1994). "Serial Cost-Sharing of Excludable Public Goods." *Review of Economic Studies* 61 (2, April): 305–25.

Moulin, Hervé and Scott Shenker (1992). "Serial Cost Sharing." *Econometrica* 60 (5, September): 1009–37.

Moynihan, Daniel Patrick (1993). *Pandaemonium: Ethnicity in International Politics.* Oxford: Oxford University Press.

Myerson, Roger (1979). "Incentive Compatibility and the Bargaining Problem." *Econometrica* 47 (1, January): 61–73.

(1991). *Game Theory: Analysis of Conflict.* Cambridge, MA: Harvard University Press.

Nash, John (1950). "The Bargaining Problem." *Econometrica* 18 (2): 155–62.

(1951). "Non-cooperative Games." *Annals of Mathematics* 54 (2, September): 286–95.

(1953). "Two-Person Cooperative Games." *Econometrica* 21 (1, January): 128–40.

Negotiating to Settlement in Divorce (1987). Clifton, NJ: Prentice-Hall.

Neyman, Jerzy (1946). "Un Théorème d'Existence." *C.R. Academie de Science Paris* 222: 843–5.

Nicolson, Nigel (ed.) (1992). *Vita and Harold: The Letters of Vita Sackville-West and Harold Nicolson.* New York: Putnam.

Nurmi, Hannu (1984). "On the Strategic Properties of Some Modern Methods of Group Decision Making." *Behavioral Science* 29 (4, October): 248–57.

(1987). *Comparing Voting Systems.* Dordrecht, Holland: D. Reidel.

Olivastro, Dominic (1992a). "Preferred Shares." *The Sciences* (March/April): 52–4.

(1992b). "Solutions and Sequelae." *The Sciences* (July/August): 55.

Olson, Mancur, Jr. (1965). *The Logic of Collective Action: Public Goods and the Theory of Groups.* Cambridge, MA: Harvard University Press.

O'Neill, Barry (1982). "A Problem of Rights Arbitration from the Talmud." *Mathematical Social Sciences* 2 (4, June): 345–71.

Osborne, Martin J. and Ariel Rubinstein (1990). *Bargaining and Markets.* San Diego, CA: Academic Press.

(1994). *A Course in Game Theory.* Cambridge, MA: MIT Press.

Oskui, Reza (n.d.). "Dirty Work Problem." Preprint, Department of Mathematics and Statistics, University of Idaho.

Passel, Peter (1994). "Economic Scene." *New York Times* (April 7): D2.

Pazner, Elisha and David Schmeidler (1974). "A Difficulty in the Concept of Fairness." *Review of Economic Studies* 41 (3, July): 441–3.

(1978). "Egalitarian Equivalent Allocations: A New Concept of Economic Equity." *Quarterly Journal of Economics* 92 (4, November): 671–87.

Pearce, David (1984). "Rationalizable Strategic Behavior and the Problem of Perfection." *Econometrica* 52 (4, July): 1029–50.

Pocock, J. G. A. (ed.) (1977). *The Political Works of James Harrington.* Cambridge: Cambridge University Press.

(1992). "Introduction." In James Harrington, *The Commonwealth of Oceana* and *A System of Politics*, ed. J. G. A. Pocock. Cambridge: Cambridge University Press.

Potthoff, Richard F. (1990). "Use of Integer Programming for Constrained Approval Voting." *Interfaces* 20 (5, September–October): 79–80.

Powell, Robert (1989). *Nuclear Deterrence Theory: The Search for Credibility*. Cambridge: Cambridge University Press.

Pratt, John Winsor and Richard Jay Zeckhauser (1990). "The Fair and Efficient Division of the Winsor Family Silver." *Management Science* 36 (11, November): 1293–301.

The Prophets (1978). Philadelphia: Jewish Publication Society.

Rabin, Matthew (1993). "Incorporating Fairness into Game Theory and Economics." *American Economic Review* 83 (5, December): 1281–302.

Raiffa, Howard (1953). "Arbitration Schemes for Generalized Two-Person Games." In Harold Kuhn and A. W. Tucker (eds.), *Contributions to the Theory of Games*, vol. I (*Annals of Mathematics Studies* 24). Princeton, NJ: Princeton University Press, pp. 361–87.

(1982). *The Art and Science of Negotiation*. Cambridge, MA: Harvard University Press.

(1985). "Post-Settlement Settlements." *Negotiation Journal* 1 (1, January): 9–12.

(1993). "The Neutral Analyst: Helping Parties to Reach Better Solutions." In Lavinia Hall (ed.), *Negotiation: Strategies for Mutual Gain*. Newbury Park, CA: Sage, pp. 14–27.

Rapoport, Amnon, Dan S. Felsenthal, and Zeev Maoz (1988a). "Microcosms and Macrocosms: Seat Allocation in Proportional Representation Systems." *Theory and Decision* 24 (1, January): 11–33.

(1988b). "Proportional Representation: An Empirical Evaluation of Single-Stage, Non-Ranked Voting Procedures." *Public Choice* 59 (2, November): 151–65.

Rapoport, Amnon and Ramzi Suleiman (1992). "Equilibrium Solutions for Resource Dilemmas." *Group Decision and Negotiation* 1 (3, November): 269–94.

Rasmusen, Eric (1994). *Games and Information: An Introduction to Game Theory* (2nd edn). Cambridge, MA: Blackwell.

Rawls, John (1971). *A Theory of Justice*. Cambridge, MA: Harvard University Press.

Ray, Depanka (1986). "On the Practical Possibility of a 'No Show Paradox' under the Single Transferable Vote." *Mathematical Social Sciences* 11 (2, April): 183–9.

Rebman, Kenneth (1979). "How to Get (at Least) a Fair Share of the Cake." In Ross Honsberger (ed.), *Mathematical Plums*. Washington, DC: Mathematical Association of America, pp. 22–37.

Riejnierse, J. H. (1994). Private communication to S. J. Brams (August 10).

Riejnierse, J. H. and J. A. M. Potters (1994). "On Finding an Envy Free Pareto Optimal Division." Preprint, Department of Mathematics, University of Nijmegen, The Netherlands.

Riker, William H. (1986). *The Art of Political Manipulation*. New Haven, CT: Yale University Press.

Riley, John G. (1988). "Ex Post Information in Auctions." *Review of Economic Studies* 55 (3, July): 409–30.

Robertson, Jack M., and William A. Webb (1991). "Minimal Number of Cuts for Fair Division." *Ars Combinatoria* 31 (June): 191–7.

(n.d.a). "Approximating Fair Division with a Limited Number of Cuts." Preprint, Department of Pure and Applied Mathematics, Washington State University.

(n.d.b.). "Asymptotic Values for the Number of Ramsey Partitions of Integers." Preprint, Department of Mathematics, Washington State University.

Roemer, John E. (1994). *Egalitarian Perspectives: Essays in Philosophical Economics*. New York: Cambridge University Press.

Roth, Alvin E. and Marilda A. Oliveira Sotomayor (1990). *Two-Sided Matching: A Study in Game-Theoretic Modeling and Analysis*. Cambridge: Cambridge University Press.

Rothkopf, Michael H. (1994). "Models of Auctions and Competitive Bidding." In S. M. Pollock *et al.* (eds.), *Handbooks in Operations Research and Management Science*, vol. VI. Amsterdam: Elsevier Science, pp. 673–99.

Rothkopf, Michael H. and Richard Engelbrecht-Wiggans (1993). "Misapplications Reviews: Getting the Model Right – The Case of Competitive Bidding." *Interfaces* 23 (3, May–June): 99–106.

Rothkopf, Michael H. and Ronald M. Harstad (1994). "Modeling Competitive Bidding: A Critical Essay." *Management Science* 40 (3, March): 364–84.

(1995). "Two Models of Bid-Taker Cheating in Vickrey Auctions." *Journal of Business* 68 (2, April): 257–67.

Rothkopf, Michael H., Thomas J. Teisberg, and Edward P. Kahn (1990). "Why Are Vickrey Auctions Rare?" *Journal of Political Economy* 98 (11, February): 94–109.

Rubinstein, Ariel (1982). "Perfect Equilibrium in a Bargaining Model." *Econometrica* 50 (1, July): 99–109.

Rumley, Dennis and Julian V. Minghi (eds.) (1991). *The Geography of Border Landscapes*. London: Routledge.

Saari, Donald G. (1994). *Geometry of Voting*. New York: Springer-Verlag.

Salovey, Peter (ed.) (1991). *The Psychology of Jealousy and Envy*. New York: Guilford.

Salter, Stephen H. (1986). "Stopping the Arms Race." *Issues in Science and Technology* 2 (2, Winter): 74–92.

Samuelson, William F. (1985). "Dividing Coastal Waters." *Journal of Conflict Resolution* 29 (1, March): 83–111.

Schelling, Thomas C. (1960). *The Strategy of Conflict*. Cambridge, MA: Harvard University Press.

Schoeck, Helmut (1969). *Envy: A Theory of Social Behaviour*, tr. by Michael Glenny and Betty Ross. New York: Harcourt, Brace, & World.

Sebenius, James K. (1992). "Negotiation Analysis: A Characterization and Review." *Management Science* 38 (1, January): 18–38.

Simon, Herbert A. (1991). *Models of My Life*. New York: Basic.

Singer, Eugene (1962). "Extension of the Classical Rule of 'Divide and Choose.'" *Southern Economic Journal* 28 (4, April): 391–4.

Sjöström, Tomas (1994). "Implementation in Undominated Nash Equilibria without Integer Games." *Games and Economic Behavior* 6 (3, May): 502–11.

Smith, H. F. Russell (1914). *Harrington and His* Oceana*: A Study of a 17th Century Utopia and Its Influence in America*. Cambridge: Cambridge University Press.

Smith, Jean Edward (1963). *The Defense of Berlin*. Baltimore: Johns Hopkins University Press.

Smith, John H. (1973). "Aggregation of Preferences with Variable Electorate." *Econometrica* 41 (6, November): 1027–41.

Smith, Vernon L. (1987). "Auctions." *The New Palgrave: A Dictionary of Economics*, vol. I. London: Macmillan, pp. 138–44.

Steinhaus, Hugo (1948). "The Problem of Fair Division." *Econometrica* 16 (1, January): 101–4.

(1949). "Sur la Division Pragmatique." *Econometrica* (supplement) 17: 315–19.

(1969). *Mathematical Snapshots*, 3rd edn. New York: Oxford University Press.

Stewart, Ian (1995). "Fair Shares For All." *New Scientist* 146 (1982, June): 42–6.

Stone, A. H. and J. W. Tukey (1942). "Generalized 'Sandwich' Theorems." *Duke Mathematical Journal* 9 (2, June): 356–9.

Straszak, Andrzej *et al.* (1993). "Computer-assisted Constrained Approval Voting." *Group Decision and Negotiation* 2 (4, November): 375–85.

Stromquist, Walter (1980). "How to Cut a Cake Fairly." *American Mathematical Monthly* 87 (8, October): 640–4.

(1981). "Addendum to 'How to Cut a Cake Fairly.'" *American Mathematical Monthly* 88 (8, October): 613–14.

Stromquist, Walter and D. R. Woodall (1985). "Sets on Which Several Measures Agree." *Journal of Mathematical Analysis and Applications* 108 (1, May 15): 241–8.

Sulzberger, C. L. (1970). *The Last of the Giants*. New York: Macmillan.

Szaniawski, Klemens (1991). "On Fair Distribution of Indivisible Goods." In Peter Geach (ed.), *Logic and Ethics*. Dordrecht, Holland: Kluwer Academic Publishers, pp. 275–88.

Tadenuma, Koichi and William Thomson (1993a). "The Fair Allocation of an Indivisible Good When Monetary Compensations are Possible." *Mathematical Social Sciences* 25 (2, February): 117–32.

(1995). "Games of Fair Division." *Games and Economic Behaviour* 9 (2, May): 191–204

Tannenbaum, Peter and Robert Arnold (1992). *Excursions in Modern Mathematics*. Englewood Cliffs, NJ: Prentice-Hall.

Taylor, Alan D. (1993). "Algorithmic Approximations of Simultaneous Equal Division." Preprint, Department of Mathematics, Union College.

(1995). *Mathematics and Politics: Strategy, Voting, Power and Proof*. New York: Springer-Verlag.

Taylor, Alan D. and William S. Zwicker (1992). "A Characterization of Weighted Voting." *Proceedings of the American Mathematical Society* 115 (4, August): 1089–94.

(1993). "Weighted Voting, Multicameral Representation, and Power." *Games and Economic Behavior* 5 (1, January): 170–81.

(1995a). "Simple Games and Magic Squares." *Journal of Combinatorial Theory*, Series A, 71 (1, July): 67–88.

(1995b). "Interval Measures of Power." *Mathematical Social Sciences*, forthcoming.

(1995c). "Quasi-Weightings,Trading, and Desirability Relations." Preprint, Department of Mathematics, Union College.

Thaler, Richard H. (1988). "Anomalies: The Winner's Curse." *Journal of Economic Perspectives* 2 (1, Winter): 191–202.

(1992). *The Winner's Curse: Paradoxes and Anomalies of Economic Life*. New York: Free Press.

Thomson, William and Hal R. Varian (1985). "Theories of Justice Based on Symmetry." In Leonid Hurwicz, David Schmeidler, and Hugo Sonnenschein (eds.), *Social Goals and Social Organization: Essays in Honor of Elisha Pazner*. New York: Cambridge University Press, pp. 107–29.

The Torah: The Five Books of Moses, 2nd edn (1967). Philadelphia: Jewish Publication Society.

Ullmann-Margalit, Edna (1977). *The Emergence of Norms*. Oxford: Oxford University Press.

Urbanik, K. (1955). "Quelques Theorémès sur les Mesures." *Fundamenta Mathematicae* 41: 150–62.

van Damme, Eric (1991). *Stability and Perfection of Nash Equilibria* (2nd edn). Heidelberg, Germany: Springer-Verlag.

Van Deemen, Adrian (1993). "Paradoxes of Voting in List Systems of Proportional Representation." *Electoral Studies* 12 (3, September): 234–41.

Varian, Hal R. (1974). "Equity, Envy, and Efficiency." *Journal of Economic Theory* 9 (1, September): 63–91.

Vickrey, William (1961). "Counterspeculation, Auctions, and Competitive Sealed Tenders." *Journal of Finance* 16 (1, March): 8–37.

Webb, William A. (1990). "A Combinatorial Algorithm to Establish a Fair Border." *European Journal of Combinatorics* 11 (3, May): 301–4.

(n.d.). "But He Got a Bigger Piece Than I Did." Preprint, Department of Pure and Applied Mathematics, Washington State University.

Weingartner, H. Martin and Bezalel Gavish (1993). "How to Settle an Estate." *Management Science* 39 (5, May): 588–601.

Weller, Dietrich (1985). "Fair Division of a Measurable Space." *Journal of Mathematical Economics* 14 (1): 5–17.

Willson, Stephen J. (1995). "Fair Division Using Linear Programming." Preprint, Department of Mathematics, Iowa State University.

Wilson, Robert B. (1992). "Strategic Analysis of Auctions." In Robert J. Aumann and Sergiu Hart (eds.), *Handbook of Game Theory*, vol. I. Amsterdam: North-Holland, pp. 227–79.

Woodall, D. R. (1980). "Dividing a Cake Fairly." *Journal of Mathematical Analysis and Applications* 78 (1, November): 233–47.

(1986a). "How Proportional is Proportional Representation?" *Mathematical Intelligencer* 8 (4): 36–46.

(1986b). "A Note on the Cake-Division Problem." *Journal of Combinatorial Theory*, Series A 42 (2, July): 300–1.

Yang, Chun-Lei (1992). "Efficient Allocation of an Individual Good – A Mechanism Design Problem under Uncertainty." Preprint, LS Wirtschaftstheorie, University of Dortmund, Germany.

Young, H. Peyton (1975). "Social Choice Scoring Functions." *SIAM Journal on Applied Mathematics* 28 (4, June): 824–38.

(1987). "On Dividing an Amount According to Individual Claims or Liabilities." *Mathematics of Operations Research* 12: 398–414.

(1994). *Equity in Theory and Practice*. Princeton, NJ: Princeton University Press.

Young, H. Peyton (ed.) (1991). *Negotiation Analysis*. Ann Arbor, MI: University of Michigan Press.

Zagare, Frank C. (1987). *The Dynamics of Deterrence*. Chicago: University of Chicago Press.

Index